D0952455

Secrets of the Snout

Secrets of the Snout

THE DOG'S INCREDIBLE NOSE

FRANK ROSELL

Foreword by Marc Bekoff *Translated by Diane Oatley*

THE UNIVERSITY OF CHICAGO PRESS

Chicago and London

The University of Chicago Press, Chicago 60637
The University of Chicago Press, Ltd., London

Published 2018
Printed in the United States of America

The translation of this work was funded by University College of Southeast Norway.

27 26 25 24 23 22 21 20 19 18   1 2 3 4 5

ISBN-13: 978-0-226-53636-1 (cloth)
ISBN-13: 978-0-226-53653-8 (e-book)
DOI: https://doi.org/10.7208/chicago/9780226536538.001.0001

Library of Congress Cataloging-in-Publication Data
Names: Rosell, Frank, 1969– author.
Title: Secrets of the snout: the dog's incredible nose / Frank Rosell; foreword by Marc Bekoff; translated by Diane Oatley.
Other titles: En nese for alt. English
Description: Chicago; London: The University of Chicago Press, 2018. | Includes bibliographical references and index.
Identifiers: LCCN 2017030476 | ISBN 9780226536361 (cloth: alk. paper) | ISBN 9780226536538 (e-book)
Subjects: LCSH: Working dogs. | Olfactory sensors.
Classification: LCC SF428.2 .R6813 2018 | DDC 636.73—dc23
LC record available at https://lccn.loc.gov/2017030476

∞ This paper meets the requirements of ANSI/NISO Z39.48-1992 (Permanence of Paper).

This book is dedicated to my mother and father, who gave Terje (my twin brother) and me our first dog, Tinka no. 1, when we were twelve years old, and to all my four-legged friends—the Tinkas, Tapio, Tapas, Shib, and Chilli.

CONTENTS

FOREWORD

Dogs have a nose for everything. If you live with a dog, or simply enjoy observing them, you know that dogs follow their noses most of the time, often in ways we wish they didn't. What secrets can a dog sniff out, and what are the secrets of the dog's olfactory talent? In these pages, Frank Rosell covers all we know about the dog's amazing nose. He shows how dogs' prodigious sense of smell has helped us in more ways than you'd imagine, and why understanding dogs' noses can make us better companions to them. He is fortunate to live with three border collies who participate in his research.

People have been interested in dog olfaction for many centuries, but only in recent decades has research in this field received more detailed attention. We are curious about dogs' noses because they are so vastly more sensitive than our own, and because we understand so little about what dogs are doing when they stop to smell a lamppost or fire hydrant. Many dog books cover some of the topics that Dr. Rosell considers, but none are as extensive, well researched, or written for a general audience.

Each chapter of *Secrets of the Snout* begins with the story of a specific dog and continues with current knowledge in the field. Throughout, Dr. Rosell shows readers how dogs' sensing abilities are helping humans. I was simply astonished at the growing range of applications for which the dog's sense of smell is being used. Through specialized training, dogs are taught to find the missing or lost, including humans and animals buried under snow or debris, and to locate an array of other items and substances, from golf balls to air pollution. Dr. Rosell also discusses how dogs help doctors by detecting diseases such as cancer and diabetes. He does an excellent job explaining what odors dogs detect and how they do it.

Dr. Rosell's stories of how dogs learn to put their noses to work for people and other animals hold many lessons for all people interested in dogs, including dog trainers. Dogs love to sniff here and there, and it's essential to allow them to do so. With a better understanding of the canine sense of smell, you can devise new and absorbing olfactory challenges for your dog. *Secrets of the Snout* offers a perspective on dogs, along with plenty of practical insight, that will greatly enrich your dog's life, and your own.

Marc Bekoff, PhD
Author of *Canine Confidential: Why Dogs Do What They Do*

PREFACE

My fascination with dogs started at the age of twelve, when my twin brother, Terje, and I got our first dog, Tinka, a Shetland sheepdog. I began to notice how mammals communicate with one another through the use of odorants. Later, at university, I focused on chemical engineering, but my fascination with the behavior of animals led me to chemical behavioral ecology as my field of expertise. I completed my doctorate in 2002 and became a professor in this subject area five years later. For twenty years I have done research on and taught about the odor-based communication of many different species of mammals, including the beaver, brown bear, yellow-bellied marmot, and European badger.

I was born and raised in Halden in Østfold County in Norway. My research has been predominantly focused on the beaver since I observed my first individual beaver on July 21, 1990, the day before my twenty-first birthday. Tinka became my constant companion on beaver excursions, and starting in the autumn of 2008, I expanded the focus of my research to include the dog. I have taken great pleasure in all the dogs I have lived with both in Halden and later in Bø in Telemark. Tinka 1 and the border collies Tinka 2, Tapio, Shib, Tapas, and Chilli have all eagerly put their noses to work, running off to chase after female dogs in heat, rolling in feces, and sniffing at other dogs' faces, behinds, and scent markings. I have often observed this with interest and wonder and, now and then, with irritation.

The dog has been man's "best friend" somewhere between 11,000 and 33,000 years, and in more recent centuries, dogs have been employed for many more purposes than solely that of a companion. This book is about dogs' sense of smell and the many ways we have put it to work. Dogs can be trained to sniff out almost everything. They have a nose for all manner of things, and there is almost no limit to the tasks they would gladly do for us. We all know about hunting dogs. Search and rescue organizations use dogs to find missing persons. Customs authorities have their own dogs to sniff out narcotics, currency, and other smuggled goods, while police dogs are trained to find weapons, blood, and semen. The armed forces train dogs to search for bombs, mines, and other explosives, while pest controllers train them to detect carpenter ants, rats, mice, and bedbugs. Thousands of dogs protect us from criminals, smugglers, terrorists, and arsonists. Dogs are also used to sniff out alien or endangered animal and plant species, to locate contaminants, and to detect diabetes and different types of cancer at a very early stage.

These are just a few of the themes that will be discussed in this book. You will also read about the wine and spirits dog Tutta, the pet finder AJ, the rescue dog Barry, the hunting dog Balder, the police dogs Trixxi and Kaos, the military working dog Lisa, the diabetes detection dogs Shirley and Nemi, the turtle dog Ridley, the Lundehunde Frøya, the orca feces detection dog Tucker, the beaver sniffer dogs Mie, Shib, and Tapas, the spruce bark beetle detection dogs Meja and Aska, the rot detection dog Cleo, the building inspector dog Luna, the oil spill dogs Jippi and Tara, the estrus detection dog Elvis, the human feces detection dogs Sable and Logan, and the golf ball dog Goya.

Enjoy the book!

Frank Rosell

1

Dogs at Work

In 1925 diphtheria broke out in Nome, Alaska, and it was vital to acquire the antitoxin serum for those afflicted with the illness. However, with winter storms and impassable roads, it seemed virtually impossible to acquire the serum before the outbreak became an epidemic. Nonetheless, dog relay teams were assembled to deliver the antitoxin, with the final leg of the treacherous journey completed by Norwegian Gunnar Kaasen. With his team of Siberian huskies led by a dog named Balto, Kaasen succeeded in delivering the serum, which prevented a deadly epidemic.

Caught in a blinding snowstorm, Kaasen had almost given up on making it to Nome with the serum until, in a final act of desperation, he appealed to Balto to find the way through the snowstorm. There was minimal visibility, so Kaasen was completely reliant on Balto's sense of smell to reach their destination. In *The Cruelest Miles*, Gay Salisbury and Laney Salisbury describe the scene:

> [Balto] understood that he had to regain the trail, to find the faint scent of the dogs that had pattered before him that winter. Balto kept his nose low to the ground, his ears flattened against his head to keep out the wind, as he moved slowly over the snow. . . . Minutes passed like hours. They were beyond the ridge and still Balto searched. Suddenly, the dog lifted his head and broke out into a run. They were back on the trail. . . . Around 5:30 AM on Monday, February 2, Kaasen could make out the outline of the cross above St. Joseph's Church. Within a few minutes he pulled up onto Front Street and stopped, exhausted, his eyes stinging from the cold, dry air, outside the door of the Miners & Merchants bank in Nome. Witnesses to this drama said they saw Kaasen stagger off the sled and stumble up to Balto, where he collapsed muttering: "Damn fine dog."[1]

Balto's life has been covered in a documentary, and there is even a statue of him in Central Park in New York. Steven Spielberg has also brought joy to many children and adults with his popular animated film about Balto's impressive achievement.

The Dog and Humans

Human beings have learned to understand and communicate with dogs through our long-lasting relationship with them.[2] We have developed more than 1,000 dog breeds, each and every one with special characteristics.[3] There are almost 900 million dogs living in households all over the world.[4] There are 75 million dog owners in the United States alone, and 40 percent of these allow their "best friends" to climb into bed with them at night.[5]

Increasingly, more scientific work is being done to analyze experiences with and opportunities for working dogs. However, the field remains underdeveloped, partly because it encompasses so many different disciplines, including agriculture, environmental studies, zoology, entomology, criminology, medicine, psychology, and wildlife biology.[6] It is my hope that this book will contribute to bringing these disparate disciplines a little closer to one another and to opening up new collaborative opportunities in the future. The dog still has a large, untapped potential as a working animal. I also hope that more dogs will have the chance to enrich their lives as working dogs, whereby they will be given a range of tasks for the use of their noses, for their own pleasure and ours. Giving dogs chances to perform work tasks and to make decisions is important for their well-being.[7]

From Wolf to Dog

When and how did the dog become our "best friend"? The Canidae family, which includes both wolves and dogs, arose 50 million years ago.[8] The dog's genome (the complete genetic material contained in a dog) was mapped out in 2003, and the results indicated that the dog stems from the gray wolf. Genetically speaking, a dog is 99.96% wolf.[9] The dog and the wolf have been viewed as belonging to the same species because they can reproduce by mating with each other and their offspring are fertile. The mating of wolves and dogs occurs most frequently between female wolves and male dogs, but can also happen between male wolves and female dogs.[10] Nonetheless, many people use the Latin name *Canis familiaris* in reference to the dog, and not the subspecies name *Canis lupus familiaris*, which others hold to be the correct term.

There is little consensus regarding when the wolf and dog went their separate ways. Many research scientists maintain that it occurred only 11,000 to 16,000 years ago.[11] Evidence has been found showing that dogs were buried together with humans 14,000 years ago,[12] which indicates that already at that time dogs were man's "best friend" and protector. In the Razboinichya Cave in Siberia, which we know was once inhabited by humans, the skull of a dog estimated to be 33,000 years old was found.[13] It was most similar to the domesticated dogs from Greenland, a breed that is approximately 1,000 years old and a variety of ancient and modern-day wolves. But this type of dog did not exist long enough to produce sufficient offspring and is therefore not the oldest ancestor of the dogs of today.

It was probably in the region that currently constitutes Germany and Switzerland that primitive humans took in the friendliest wolves as a means of protecting themselves from cave bears and lions. This implies that the taming of dogs first occurred in Europe and not in Asia, as was formerly believed. These findings from 2013 indicate that the domestication of wolves took place as far back as 18,800 to 32,100 years ago, when large parts of northern Europe were still covered by ice. When some of the friendlier wolves began slinking around the camps of these ancient civilizations in search of mammoth flesh, they were welcomed, since they served as watchdogs. Over time with their acceptance into human society, the wolves began eating food that contained more vegetable starch (formed in most green plants).[14] An alternative possibility is that humans sought out wolf dens and captured and tamed the wolf pups.[15]

A 2013 study led by Erik Axelsson—a scientist in evolutionary genetics at the University of Uppsala in Sweden—found thirty-six specific areas in which the genomes of the dog and the wolf are different. Nineteen of these areas contain genes involved in brain development, which could explain why dogs are friendlier than wolves. The researchers also discovered that dogs have ten genes that help them to digest starch and break down fat. Three of these genes make dogs better equipped than wolves to break starch down into sugar, so it can be absorbed.[16] Most dogs are raised by humans, and their diet can play a very important role in their food preferences later in life. Unlike adult dogs, puppies have a clear preference for meat.[17] We also know from epigenetics (the study of heritable changes in gene expression and how the genes are employed) that offspring are influenced by the experiences of their parents. For example, a laboratory mouse that has been trained to avoid certain smells could pass this learned behavioral trait on to its offspring. This is called epigenetic DNA programming, whereby the genes can be switched on or off.[18] More of this type of research is being done and will potentially contribute to explaining the large differences we find between wolves and dogs, including behavioral differences.

In 2014 postdoctoral student Adam H. Freedman and colleagues at the University of California, Los Angeles, analyzed the genomes of gray wolves from three locations (China, Croatia, and Israel) where the domestication of the gray wolf may have occurred. They also studied the genome of a basenji from Africa and that of an Australian dingo. Both of these dog breeds come from areas without gray wolves, and therefore they could not have at any time mated with gray wolves. The researchers found that the gray wolves from the three locations had more in common with one another than with dogs. They also studied the genome of a boxer and discovered that the dog breeds from the three respective locations had more in common with one another than with the gray wolves. This indicates that modern-day dogs and gray wolves represent sister branches on the evolutionary tree and that they both stem from an older, now extinct, common ancestor. These findings are inconsistent with earlier speculations that the dog evolved from one of the three gray wolf populations.[19]

Some wild dog packs have a dominance hierarchy, in which individual dogs will have advantages related to food and mating, but this is not as pronounced as in wolves. In the case of wild dogs, it is not an alpha pair that leads the pack; instead, the leader is usually an older and high-ranking individual dog. High-ranking dogs who are met with appeasement behavior in both greeting ceremonies and in hostile contexts more frequently lead the pack than dominant dogs who are greeted with appeasement behavior only in hostile situations. In other words, dominant dogs are those with the largest number of friendly relationships, and the friendliest dog of all is often the leader of the pack.[20] Whether a dog wins or loses a game of tug-of-war will not make it more or less dominant in relation to its owner.[21] Dogs prefer not to challenge higher-ranking pack members. This trait is what helps us to have control over and handle our dogs.[22]

The Nose at Work

The dog has a very keen sense of smell, which has been used in the service of humans for many thousands of years. In general, the dog's nose is 100,000 to a million times more sensitive than the human nose.[23] The dog's rhinencephalon (smell-brain) is almost seven times larger than that of humans, and with their fantastic sense of smell, dogs are able to perform many work tasks for us.[24] It all began when humans put the dog's nose to good use for hunting. We continue to discover the ways that dogs' noses can be used to help us. Dogs have been used in wars, not only as protectors but also to find explosives and land mines following a war. In the course of the past forty years, the use of specially trained sniffer dogs has increased dramatically. These dogs typically search for odors from human beings or the particular odor emitted by a specific object. Search and rescue dog organizations use dogs to search for missing persons; customs authorities have dogs specially trained to detect narcotics, cash, and other smuggled goods; and police dogs are trained to find weapons, blood, and semen. The armed forces use dogs trained to search for bombs, mines, and other explosives, while pest control companies have dogs trained in the detection of carpenter ants, rats, mice, and bedbugs. In short, thousands of dogs protect us from criminals, smugglers, terrorists, arsonists, and pests.[25] Dogs are also used to sniff out alien or endangered animal and plant species, to locate contaminants, and to detect diabetes and individual types of cancer at very early stages.

Dogs can be trained to sniff out just about anything, and our imagination is virtually the only limit when it comes to potential work tasks for canines. The most important thing is that dogs can be trained to communicate to us the information they acquire by using their noses. The dog is the most successful mammal on earth after human beings, and one of the reasons for this is that they are very willing pupils. Examples of their unique learning capacity are found in the 2013 book *The Genius of Dogs* by Brian Hare and Vanessa Wood.[26] The stories in the book about the border collies Chaser and Rico illustrate dogs' learning abilities and their potential brain capacity.[27] Chaser was born in May 2004, and when she was five months old, John W. Pilley, a retired psychology professor at Wofford College in South Carolina, started teaching her different words. Over the next three years, Chaser learned and remembered the names of 1,022 different objects. These included everything from stuffed animals and balls to Frisbees and different plastic objects. In the course of 145 tests using 20 objects in each test, Chaser identified in all cases a minimum of 18 out of 20 objects (approximately 90 percent correct). In another test, Chaser was trained to pick up an object with her mouth, move the object with her front paw, or touch the object with her mouth or nose. For example, when Pilley ordered her to "pick up Lambs," she was supposed to pick up the stuffed animal Lambs with her mouth. She was given fourteen similar work tasks and performed all of them correctly.

Chaser also knows that different objects can be one of many in a category. For example, "ball" is a category containing 116 round and bouncing objects. She could also find an unfamiliar object by logically eliminating other potential alternatives. She managed in eight successive repetitions to retrieve an object she had never learned the name of because this object was grouped together with otherwise familiar objects. Twenty-four hours later, however, she had forgotten the name of these new unfamiliar objects. For Chaser to develop a long-term memory of unfamiliar objects, an exercise involving repetition is required.[28] In a final test, Chaser was given the command "to ball take Frisbee" followed by "to Frisbee take ball." She understood which object was to be brought to the other in 78 percent of the cases when a number of familiar objects were used in a sentence.[29] Her training ended after three years, not because the limit for Chaser's learning capacity had been reached, but because Pilley could no longer spend four to five hours a day training her.[30] Chaser learned our language in exactly the same way as a three- or four-year-old child would. Most of the words she knew could be used in different contexts and in new sentences without the need for additional learning.[31]

Through the domestication of dogs, we have developed a unique bond with them. If we have one dish that smells of food but point at another dish, the dog will not use its sense of smell; it will go in the direction we are pointing instead. This shows how much they trust us.[32] Less surprising is the fact that a dog trusts its owner more than strangers.[33] Without training and socialization, dogs are actually better than wolves and chimpanzees at understanding our hand gestures, although chimpanzees are smarter than dogs in most other situations. Working dogs are the most intelligent of all when it comes to reading our movements.[34] They are extremely motivated when it comes to carrying out a task correctly, even if they do not receive an immediate reward.[35] When it comes to determining the best dog breed, there is no scientific evidence demonstrating that one breed is smarter than another.[36] The most common work dogs are German shepherds, Belgian sheepdogs (Malinois), English springer spaniels, Labrador retrievers, golden retrievers, and border collies. These breeds are intelligent, strong, loyal, impressionable, and, above all, willing to learn.

Dogs are able to adapt to fluctuating work hours because they have a naturally short sleeping pattern with frequent sleep-awake cycles.[37] Still, it is important to remember that dogs also need to take breaks when working and that four hours of work a day is a good rule of thumb.[38]

It is not only scientists and those who use dogs in their work on a daily basis who are interested in work tasks for the dog's nose. Nose work is becoming popular in many different communities internationally.[39] Courses in the specialized training of sniffer dogs are becoming more and more common, both in a professional capacity as well as within the private pet market.[40] It is both physically and mentally stimulating for dogs to use their noses, and it is an activity that is good for all dogs. For example, a dog can easily be taught to search for treats, to find different objects (toys or things we have lost, such as car keys), and to follow different trails (of a pancake, a hot dog, or human).[41] The video *Nose Work* describes these search games in detail.[42]

Dogs can carry out searches in laboratories and other locations.[43] There are many ways to organize scent-detection training for dogs.

The dog can either be transported to a specific site to carry out a room search, small terrain search, or field search, or a scent sample can be transported to the dog when it is working in the field. Scent samples can also be transported to a dog in the laboratory. The dog becomes a kind of detector and sometimes can be even more effective than an analysis instrument. When a scent sample is transported to a dog in a laboratory setting, the dog is presented with a multiple-choice method. Over time, a number of devices have been developed, each requiring its own search methods, such as a labyrinth, scent discrimination box training/box training apparatus, training platforms, scent detection boards, and a training wheel/carousel. The last three are the most common. Originally, cans were attached to chairs that could be moved around; later the use of a round table was implemented.[44] This multiple-choice method was developed for the first time in the 1960s[45] and is used for the specialized training of many types of sniffer dogs, such as those used for tracking semen, blood, explosives, mines, mushrooms, and environmental toxins.

Research with My Own and Others' Beaver Sniffer Dogs

In 2001, in the course of working on my doctorate, Lars Joran Sundsdal (my master's degree student at that time) and I discovered that during the winter beavers deposit a substance called castoreum in their scent mounds.[46] Little is known about the beaver's anal gland secretion, but we do know that it is deposited in the beaver's scent markings throughout the spring and summer.[47] Castoreum contains no detailed information about the individual and merely states, "I live here; this is my territory." The anal gland secretion contains information about sex, but we don't know if this is also true for castoreum. When we performed chemical analyses of castoreum on a gas chromatograph using a mass spectrometer, we found no differences between the sexes.[48] I therefore developed an interest in investigating whether dogs would be able to distinguish between the sexes using their sense of smell and also if they would manage to differentiate castoreum from anal gland secretion. In the spring of 2013, associate professor Andreas Zedrosser, PhD candidates Hannah B. Cross and Helga Veronica Tinnesand, and I gave dog trainer Tor Iljar from the company Dogpoint responsibility for training eight dogs to distinguish between castoreum and anal gland secretion from beavers. They were also trained to differentiate between the castoreum of males and females and the anal gland secretions of males and females. Iljar was responsible for the Labrador retrievers Demi and Andrea. Dog handler Marit Sorum with the Jack Russell terriers Petra and Ronja, Mia Palmgren with the poodle Zappa, and Nina Hansen with the papillon Mie and the border collies Vims and Liz also took part in the project. The dogs were between two and seven years old. At the start of a trial, it is important that a dog's owner be present, since this will increase the dog's motivation.[49] The dogs also find it easier to interpret their owners' positive states of mind than those of strangers.[50]

There are many variations among dogs and dog trainers, so it is difficult to follow a single set rule for training.[51] Countless books have been written about how to train a dog, and correspondingly many methods have been devised. When the five best-selling books about dog training—among these, the book written by the well-known dog trainer Cesar Millan[52]—were summarized, the contents proved to be highly divergent. In 2012 PhD candidate in psychology Clare M. Browne at the University of Waikato in New Zealand and her colleagues concluded that the books did not necessarily have the information required to enable dog owners to learn how to carry out a training task.[53] This confirms an old joke about how if there are four dog trainers in a room, there will be five different opinions. Unfortunately, scientists have shown very little interest in studying and comparing different training methods.[54]

In our study, the first training phase involved introducing the dogs to a training platform containing seven holes, with a can in each hole. Four of the holes were at all times inside the training platform's two Plexiglas walls, and three of the holes were outside. These holes containing cans could be moved to different positions using a handle so all the cans could be situated inside the Plexiglas. The handle was attached to the hole for can number four. Iljar used tea as the scent the dogs were to search for. A cotton ball was dipped in tea and placed in the can with the clicker handle attached. The handle is always behind the correct scent sample during the training session. At the start of the training sessions, the scent samples were put in cans made of stainless steel to reduce the amount of uncontrolled scent transmission between the cans. When the dog handler said, “Find the scent,” the dogs started to sniff at the holes containing the cans, and when they found the right scent, the dog handlers confirmed this with the clicker (so the dogs would associate a correct indication of the target with this sound), before immediately rewarding them with a treat. A positive outcome—a correct indication—will in this way be a good experience for the dog, and it will thereby be motivated to do another search. The handle was moved to different positions, and again the dogs had to find the tea scent. When they succeeded, clean cotton balls were placed in the six other cans. Other distracting scents, such as human and food odors, were also introduced, so the dogs would learn to ignore these and search through all the cans to find the correct scent. When the dogs had achieved twenty correct indications and maintained each indication for more than five seconds, this phase was finished. All of the dogs managed this in the course of seven sessions with one to two hours of training per session.

It is important to vary the scent samples in the course of the training. At Telemark University College (TUC), we have many samples from both live-captured and shot Eurasian beavers. The samples are stored at –20°C. Many scientific studies done in the 1980s and 1990s produced poor results. It turned out that the training materials were contaminated and the dogs reacted to everything from the odor of the person handling the material to the tape and Magic Marker ink on the samples. There are also reports showing that dogs have only reacted to the scents the trainer has used and not to other types of scents.[55] Dogs have a good memory for scents and can remember scent samples used in previous training sessions. It is therefore important to use one set of scent samples during the training session and completely different and unfamiliar samples in the experiment. It is also important to use many different samples. A common problem with scent samples used during training arises when these are stored together with other materials, because they can absorb odors.[56] After a training set has been used for a while, it will become contaminated and should be replaced. It is recommended that the samples be stored in a glass jar with a Teflon lid rather than in plastic bags. It is very important to use different sample materials.

In phase two of our study, the dogs were trained on castoreum (four dogs) or anal gland secretion (four dogs) from males or females. The beaver scent was presented on cotton balls. The six other cans contained unscented cotton balls. The other sex was subsequently introduced and also the other type of scent from the same sex, so the dogs had to distinguish between the scents. For example, the papillon Mie was first trained using solely the anal gland secretion of males. Subsequently, she was introduced to the anal gland secretion from females and castoreum from males on the training platform.

Dogs are very good at "reading" their handlers and thereby interpreting cues from the handler regarding which can (hole) is the right one. The dog handlers often send signals without being aware of it. Corresponding research errors were discovered in connection with the horse named Clever Hans in the early twentieth century. It was claimed that Clever Hans could count and perform other cognitive tasks. Could he really? Many believed so, until 1911 when the psychologist Oskar Pfungst discovered that the horse was responding to very small, unintended postures and facial expressions from his owner, math teacher Wilhelm von Osten from Germany, and from members of the audience.[57] German painter Emilio Rendich also doubted the horse's abilities. He therefore trained his dog, Nora, to master the same type of reaction pattern. Nora was supposed to bark as many times as her trainer wanted. When she had done this, Rendich leaned forward and Nora stopped barking.[58]

Dogs are extremely obedient. In a 2003 scientific experiment carried out by the scientist Viktoria Szetei and her colleagues at Eötvös Loránd University in Budapest, it was demonstrated that approximately half of the dogs would go to an empty food dish because the owner pointed at this dish. The dog ignored the other dish, even though it contained and smelled of food.[59] Lisa Lit, a professor at the University of California, Davis, led a study in 2011 that investigated whether the dog handler's body language influenced the dog's results during a room search.[60] The dog handlers were tricked into believing there were drugs and gunpowder odors in different rooms in a church. Any indications they might give would therefore be inaccurate. Eighteen dogs were used, and they gave 225 incorrect responses. There were only 15 percent clean runs and 85 percent runs with one or more alerts. In other words, the handlers' assumption that there was a scent present influenced the dogs. The dogs were responding to unintentional cues from the handlers. Dogs can react to pointing, staring, and head movements in the direction of the target,[61] and it is best to train dogs so they are unable to read the dog handler's body language.[62]

When the dog has advanced to a certain level, blind tests must be carried out to ensure that the dog and handler are on the right track. The story of the dog Nora shows how important it is to carry out such tests "blind," because dogs can read our body language. And since the dog handler doesn't know in which can the scent is to be found, this prevents any unconscious influence on the outcome. The person who puts out the scent—the test leader—must not speak with the dog handler or be found in the same room as the equipage (dog and handler), so as to prevent this person from influencing the results. All our trials were filmed by three video cameras, and the test leader Hannah B. Cross observed the trials on an on-camera monitor in another room.

Dogs have better eyesight than was formerly believed,[63] even though they have difficulties distinguishing between red and green. They find it easier to detect the colors light blue, gray, and yellow.[64] In a scent detection task, it is therefore important that it be difficult for the dogs to see the target, in order to ensure that they are using their sense of smell rather than their eyesight. Dogs are also sensitive to ultraviolet light, which increases the visibility of urine markings. This kind of sensitivity is normally found in species that are partially nocturnal.[65]

After Mie and the other dogs became proficient at distinguishing between the two types of odors and also between the different odors of the sexes—in other words, at finding the correct odor—we were ready to carry out a scientific experiment. In general, scientists apply "sensitivity" and "specificity" in the interpretation of results.[66] Sensitivity represents how often a dog gives an indication for a can containing a beaver scent, for example, when there is in fact a beaver scent in this can; while specificity represents how often a dog refrains from giving an indication for a can without a beaver scent. Or to simplify this a bit: how many times the dog manages to find the right scent/can and simultaneously avoids incorrect indications for the wrong odor. The goal of all the training is for the dog to become accurate. It is said that it takes at least three weeks of training to achieve a 90 percent correct result, three months to reach 90–95 percent correct results, and three years to advance to 95–100 percent correct indications.[67] This, of course, depends on the scents the dogs are supposed to recognize, the breed of dog, the training they have received, the method (the number of cans—the more cans, the more difficult for the dog[68]), the particular dog being trained, and the dog handler.[69]

In the "blind" beaver experiments, only four of the cans were used and four drops of beaver scent were put on the cotton balls. Four different scents were put in the four different cans. The correct scent was put in the first can, in the second can the same scent type but from the other sex, in the third can the other scent type from the original sex, and in the fourth can a clean cotton ball. These scents were randomly placed in one of the four cans for each trial. The training platform was cleaned with vinegar between each trial. We also carried out similar trials using scent markings from known male and female beavers that we had gathered in the field. It turned out that the dogs were able to distinguish between castoreum and anal gland secretion (100 percent correct for all the dogs with the exception of one dog that made one mistake) and between the sexes for both castoreum (92 percent) and anal gland secretion (88 percent)—for anal gland secretion, three out of the four dogs had 100 percent correct results; the dog that was unsuccessful had stepped in for a dog that was ill and thus had not received sufficient training. For castoreum, the dogs proved to be more accurate than our gas chromatograph.

We have not had the opportunity to take dogs with us out into the field. If the dogs are equally adept out in the field as they have been in the laboratory, they will provide us with many answers to questions about how the beaver defends its territory, something that has been difficult to establish using other methods. We can, for example, find out where the males and females, respectively, leave territorial markings and whether they do so with castoreum and/or anal gland secretion.

In the autumn of 2013, I started training my dogs with other dogs, in collaboration with PhD candidate Hannah B. Cross, master's degree student Christin Beate Johnsen, and dog enthusiast Beate Jaspers. I assumed responsibility for my own border collies, the siblings Chilli and Tapas (born in 2007), and their mother, Shib (born in 2005). Christin was responsible for Bailey (born in 2010), the Nova Scotia duck tolling retriever (also called a Toller), while Beate took responsibility for the Samoyed dogs Danny (born in 2009) and Shanie (born in 2003), and the papillons Triana (born in 2011) and Carmelita (born in 2011). In Finland there are both North American and Eurasian beavers. It is very difficult to distinguish between the two species by appearance and behavior, but it is possible using chemical analyses in a gas chromatograph or by genetic analyses.[70] In the north of Finland, there are two small populations of approximately fifty North American beavers, which can potentially spread to both Sweden and Norway. We don't want the North American beaver in Norway or Sweden, since it is an alien species and can therefore oust our own beaver species, the Eurasian beaver.[71] I suggested that we should start training our eight dogs, twice a week, to differentiate between the castoreum scents of the two beaver types, since this is what is most frequently deposited in scent markings out in the field.

I started training my own dogs at home in my living room. I asked the others to do the same with their dogs. I placed a treat in one of three plastic cups that I put on the living room floor, spaced at approximately 10-centimeter intervals. Each time the dogs gave an indication for the plastic cup containing the treat, I confirmed this using a clicker and gave them verbal praise. They also received the treat in the cup. It did not take long before they all did this correctly every single time, in ten successive trials. This was also a nice start for teaching them the "down" indication. In 2014 the Swedish researcher Ragen T. S. McGowan at the Swedish Agricultural University and her colleagues demonstrated that dogs like to solve tasks and that they can self-regulate access to a reward.[72] The beagles in their study became happier when they had a "eureka" moment. That was exactly how I felt when I saw my dogs solve the work task they were given. They had become very proficient in a short period of time. The researchers also found that the dogs were happier when they had earned a reward by performing a work task than when they merely received the reward. The dogs also preferred a reward of food rather than petting.

The next step of the training was to introduce the dogs to the training platform. We started with some treats in one of the cans. When they had learned to respond correctly to the treat, we introduced them to the beaver scent. We initially used the scents of dead animals that had been in the freezer since the end of the 1990s. We did so because we wanted the dogs to be trained to detect the heavier compounds. The light, volatile compounds disappear when the samples are stored for a long period of time. The dogs Tapas, Danny, Chilli, and Triana were trained to recognize the Eurasian beaver scent, and Bailey, Shib, Carmelita, and Shanie were trained to recognize the North American beaver scent. We started working using only the correct beaver scent. When the dogs had learned to recognize this scent, we introduced the scent from the other type of beaver. Now they had to distinguish between the two types of scents. We also trained the dogs to ignore scents from other animals, such as moose, roe deer, and red deer, which are found around the beaver's habitat. After the dogs had understood the work task on the training platform, we started training them to keep their backs to us—so they wouldn't pick up on any unintentional cues from us.

We later switched to using a scent detection board with six holes containing cans. This number was used the most often, so we could put out the scent samples randomly, based on a throw of a dice. Normally, only one of the cans contains the scent the dogs are searching for. The five others were used as control samples.

All eight dogs clearly enjoyed the training sessions. When we let them out of the car, they ran ahead of us through the garage and over to the training room at TUC and began wagging their tails eagerly.[73] When the dogs enter the room where they are going to perform a search on a platform or a scent detection board, they wag their tails to the right, and when they indicate the correct sample, they will wag to the right even more (see pp. 21–22 for interpreting tail wagging). They are, in other words, in a positive state of mind and demonstrate beyond any doubt how much they enjoy this work.[74] We also have outdoor training sessions. Here there are many more sounds and movements that can cause the dogs to lose their concentration. We have horses in our neighborhood, and Shib in particular was distracted by them. The wind can also represent a challenge. If it blows gently toward the dogs, they will often skip a can and go directly to the correct sample. If the wind is too strong, they can find it more challenging to find the right scent.

The training results thus far show that the dogs are able to distinguish between the two types of castoreum. We also tested scent markings from familiar Eurasian beavers from the region of our study. They also recognized these scent samples, even though they were completely fresh. Tapas, Bailey, Danny, Shib, and Carmelita learned to differentiate between the two beaver types the most quickly; the other dogs needed more time to achieve a stable outcome. Chilli was very up and down and had more bad days than the others. If somebody started throwing a ball for her or just said the word "ball," she became more interested in looking for the ball. The personalities of the dogs are slightly different, and they therefore react a bit differently when they find the right beaver scent. Chilli lunges at the correct scent with her nose directly over the can, while Bailey "digs" furiously before giving a down indication. Tapas has a more relaxed down indication style, while Shib pushes with her nose and may also lick a little bit at the hole with the can containing the correct scent.

In a blind pilot experiment we carried out using the platform with four of the dogs in the spring of 2014, Chilli and Shib gave correct indications in 95 percent of the cases. Tapas and Bailey, on the other hand, were not themselves because Shib was in heat. They were more interested in sniffing at places where she had been moving around than they were in the beaver samples. Later, all eight dogs have shown that in roughly 95 percent of the cases, they are able to find the right beaver scent in blind tests. However, we must carry out more experiments with scent markings from Finland before we can publish our results in a scientific journal, because we need more trials from the North American beaver. We have allied ourselves with the research scientist Janne Sundell from the University of Helsinki. In the autumn of 2014, Sundell gathered scent-marking samples from both types of beavers in Finland. We also wanted to investigate the scent markings using genetic methods, so we would be completely certain of which species we were dealing with before testing them on the dogs.

If the wildlife management authorities or scientists suspect that there are North American beavers in their region, the beavers' scent markings out in the field can be gathered and sent to us. Sample collection can be done by transferring the portion of the marking containing the castoreum scent to a jar or plastic bag. Then we can have our dogs determine on the training platform or scent detection board whether the marking is from the North American or Eurasian beaver. If the scent markings are from the North American beaver, the dogs trained on this type will give an indication for it, and not for the others, or vice versa if the scent marking is from the Eurasian beaver. We can then determine which type of beaver is living in areas where the two types are found (besides in Finland, both types are found in Germany, Belgium, Luxembourg, and Russia), without having to capture the animals.

Choosing the Right Working Dog

It is important to train the right dogs to become working dogs. Dog experts have clear ideas about what constitutes a good working dog even though there is little documented evidence. In 2004 the English animal welfare and behavioral biologist Nicola J. Rooney at the University of Bristol asked 244 dog trainers in Great Britain about their ideas/thoughts regarding the features of a good working dog. Based on their answers, she established that in assessing a dog's suitability, the ten most important characteristics were as follows:[75]

1. Acuity of sense of smell
2. Ability to find an object located out of sight
3. Health
4. Tendency to hunt using only sense of smell
5. Stamina
6. Ability to learn from being rewarded
7. Tendency to become distracted when searching
8. Agility
9. Consistent behavior from one day to the next
10. Motivation to search for an object

Ideally, the dog should score high in all of these areas, with the exception of item 7. According to Rooney's studies, the English springer spaniel was the breed most often used for specialized search and detection purposes in Great Britain, followed by the Labrador retriever and border collie. The abilities of males and females as specially trained sniffer dogs were very similar, but different in one sense: the males were more aggressive toward other dogs.[76]

Through the use of simple methods, it is possible to test whether your dog favors the right leg, left eye, and how it performs in a jumping test, in order to establish whether it has the potential to be a good working dog. PhD candidate Lisa M. Tomkins, who works at the University of Sydney, has found that the best seeing-eye dogs favor the right leg and the left eye and also had greater hind-leg clearance in a jumping test. The best dogs also had chest cowlicks that swirled counterclockwise.[77]

The Dog's Personality

The makeup of the dog's personality stems from a mixture of genetic and environmental factors.[78] Playfulness, curiosity, fearlessness, sociability, and aggression are characteristics of the dog's personality.[79] Other characteristics are extroversion, motivation, a focus on training, friendliness, and nervousness.[80] In 2013 behavioral biologist Erika Mirko and colleagues at Eötvös Loránd University in Budapest evaluated a series of different types of behavior to test dogs' aptitudes as working dogs.[81] There are a number of considerations to keep in mind, in that a dog can be influenced by very many different factors throughout its lifetime. If a German shepherd puppy is exposed to a number of specific environmental factors early in life, this can have long-term effects on its behavior.[82] Both the mother's character, the number of siblings in the litter, the sex of the siblings, and time of year when the puppy is born will all have an impact on how it copes with stress as an adult. The puppy's sex and body weight are also important.[83]

There are three important developmental stages in the life of a dog that can explain, in part, why dogs differ. The first, the *stimulation stage*, lasts from the time the puppy is 3 days old until it is 16 days old. The second stage, called the *socialization stage*, lasts from when the puppy is 2–3 weeks old until the age of 12–14 weeks. The third is the *enrichment stage* and lasts until the puppy is one year old. If the puppy is exposed to what is called mild stress—such as handling, playing, and petting—during the first period, it will cope with stress better as an adult. During the second stage, it is important to take the dog away from its place of birth to enable it to meet strangers. In this way, one prevents shyness and general restlessness in the dog when in the company of strangers. In the third stage, it is important for the dog to have a broad range of interesting, new, and exciting experiences. The dog should be permitted to freely explore and touch different objects and have social contact with human beings and other species. An under-stimulated dog will be anxious around new objects and have the tendency to retreat from rather than explore new situations.[84]

The Dog's Mind-Set

Both when meeting a working dog and with dogs in general, you should be attentive to the dog's mind-set. When dogs meet in public spaces, they sniff each other. Males sniff females more often, and they also mark the most. Dogs that are not on a leash sniff each other more frequently than dogs on a leash. When dogs are on a leash, they will demonstrate threatening behavior twice as often. This can be a sign that dogs become frustrated when they are not allowed to greet one another.[85] The owner's personality, attitude, and sex can also influence the dog's behavior.[86] Professor Petr Řezáč at Mendel University in Brno in the Czech Republic and his colleagues discovered that the sex of the dog's companion had the greatest impact on whether a dog would threaten or bite another dog. When the dog was being walked by a man, their aggression increased, and the research scientists thereby concluded that this could be because dogs imitate the emotions of their companions. If owners behave protectively or with self-confidence, it is likely that their dogs understand this.[87] Male dogs owned by women were less social when meeting other people. This can be an indication that male dogs develop another social role when their owners are women, since women tend to have a more relaxed relationship with their dog than male owners. Dogs that were with more self-confident male owners would often assume a subordinate role. When with women, in some contexts, the dogs would assume a dominant role.[88] Border collie puppies, who score high on sociability (how much time passes before the puppy makes contact, the amount of tail wagging, jumping, and appeasement), when meeting a stranger more frequently choose to actively seek a conflict resolution strategy. Those who scored low on sociability responded more passively.[89]

Dogs and Tail Wagging

Tail wagging can tell us a lot about a dog's frame of mind. Wagging is not just wagging. The idea that when dogs wag their tails they are happy and friendly is perhaps the most common misinterpretation of dogs. Some tail wagging is without a doubt a sign that a dog is happy. Other kinds of tail wagging, however, can mean that the dog is afraid or insecure. It can even be a warning that if you approach the dog, it will bite you. The position of the tail, the movement pattern, and speed are all of great significance. The height at which the dog holds its tail communicates something about its emotional state. When the tail stands straight up in meeting with another dog, this sends a clearly dominant signal (I'm the boss here) or a warning signal (go away or deal with the consequences). If the tail is under the body, the dog is frightened (please don't hurt me). At the same time, it is important to be aware that different breeds hold their tails differently by nature, so this should be taken into consideration when interpreting their mood. In addition to this, the movement of the tail sends another signal. If the dog wags its tail energetically, by swinging its hips from side to side, it is clearly showing that it is very pleased to see you again. If the tail wagging has rapid and short movements (vibration) and the tail is simultaneously held high in the air, it can mean that the dog is ready for a fight (an active threat signal). It is important to notice whether the dog is wagging to the right or the left. If they have positive feelings (left hemisphere activation) about something or someone, they wag more to the right side of their hindquarters (to the left if the dog is facing you, in other words, viewed from in front), while they will wag more to the left (right hemisphere activation) when they have negative feelings. In 2007 Professor Angelo Quaranta and colleagues at University of Bari Aldo Moro in Italy showed that if the dogs could see their owner, they wagged their tail more to the right. If the dogs spotted a stranger, they also wagged their tail to the right, but not as far out. They wagged even less to the right if they saw a cat. If the dogs saw an unfamiliar, dominant dog or if they were alone, they wagged their tails to the left.[90] Dogs watching a video of a dog wagging to the left, show higher brain activity and become more anxious than if they see a dog wagging to the right.[91]

The 10 most popular dog breeds of the world in 2013:[92]

1. Labrador Retriever
2. German shepherd
3. Poodle (all sizes)
4. Chihuahua
5. Golden retriever
6. Yorkshire terrier
7. Dachshund
8. Beagle
9. Boxer
10. Miniature Schnauzer

2

A Dog's Sense of Smell

It is said that a dog that has lost its sense of smell is no longer a dog. The dog trainer Torun Thomassen had the chance to experience this firsthand.[1] She had a German shepherd puppy, Ometyst's Arthur, who was born without a sense of smell. This dog was incapable of putting its nose to work on anything productive. He was the largest of the litter at birth, but unlike his siblings, he lost weight during the first days of his life, because he was unable to sniff his way to the teats of his mother to nurse. However, he did learn an alternative method after a few days. He would lie down on top of another puppy, wait until it let go, and then latch on to his mother's teat. At home, he often stole raw onions or oranges, and with time, as the entire litter matured enough to go for walks together, Arthur would inevitably be the one who ate something dangerous and got sick, or the one Thomassen had to search for. The first time he was out and met many people (many legs), he considered what he should do when there were so many legs to choose between or when some of the legs disappeared. He was not able to sniff his way back to Thomassen's leg. The legs he chose to follow turned out not to be hers.

The first time he went along on a bike ride, he experienced further problems. He was running loose, while his mother was on a leash hitched to the bicycle. They were standing at the top of a hill when two people came walking up the hill. Thomassen waited and said nothing, just to see what Arthur would do. He looked at the bicycle, his mother, and Thomassen, and subsequently at the two pairs of legs that were approaching. After having considered and reconsidered several times, he chose the legs and went along with them. Thomassen called him back, and he never made that mistake again. Instead of sniffing other dogs, he stood immobilized like a statue while they sniffed him. When they had finished, he walked away from them.

Arthur did not function socially with strange dogs. He couldn't smell anything and was therefore unable to communicate with them correctly. He was often in the company of other puppies of different breeds while training forest drills, such as tracking exercises and field searches. And even though Arthur was enthusiastic, he only found pieces of hot dogs if he happened to stumble upon them. However, he was extremely skilled in obedience and loved being trained. There were not any exciting scents to distract him during the training sessions. Arthur was always a cheerful pup—until the day he was attacked by another dog. After this, he became unpredictable and might attack his mother, for example, without warning. He started lunging at dogs who wanted to sniff him, often adult male dogs. Thomassen was therefore, unfortunately, obliged to have him put down when he was about one year old.[2] He was a wonderful dog and Thomassen was very fond of him, but she couldn't take the chance that he might hurt somebody, and she also had to take Arthur's mother into consideration.

The Development of the Sense of Smell

The sense of smell is much less evolved in human beings than it is in dogs. This makes it more difficult for us to understand and appreciate the dog's fantastic sense of smell. We can't see odors. The dog's sense of smell has evolved over the course of many thousands of years of natural selection, ensuring that the dog has the best possible adaptation to the environment in which it lives. The dog's sense of smell is important for finding food, for reproduction, for recognition of kin, and for identifying dangerous situations. Dogs with the best sense of smell have passed on their genes for generations. The result is an extremely well-developed olfactory system that is capable of discovering (or detecting) and distinguishing between different odors. The dog has an incredibly keen sense of smell, which enables it to perceive a large amount of different kinds of odor-based information.[3] The dog's nose is much better developed for detecting odors than our nose. When the dog's nose is wet and cold, from glands that produce an oily type of fluid,[4] it is easier for them to detect odors. If its nose is dry, the dog will moisten it with its tongue.

The dog's sense of smell develops in the course of the first two weeks of its life,[5] but it wasn't until 2006 that we learned that puppies can also learn scents before they are born. In the course of a series of experiments in 2006, research scientist Deborah Wells and Professor Peter Hepper at Queen's University in Belfast, Northern Ireland, studied dogs' ability to learn scents through the mother's diet while they were still in the womb. If the mother was given aniseeds, puppies that were 24 hours old preferred this scent to a greater extent than puppies that had not been exposed to this scent in the womb. They accordingly preferred what they had experienced before birth. Puppies that were tested 15 minutes after birth also showed a similar preference for aniseeds. Scent can thus be learned before birth.[6] In another experiment the same year and by the same research scientists, puppies were given aniseeds while they were in the womb and again immediately after birth. These puppies also preferred the aniseeds.[7] This ability to learn is an important adaptive trait that ensures the puppies' development and survival. It is important for the puppies to be able to recognize and become attached to the mother, and whatever she has eaten is therefore also safe for them to eat.[8]

In order to understand how the dog manages to carry out different types of work tasks, one must understand how the nose functions.

Dogs have two important olfactory organs: the *olfactory system* and the *vomeronasal system*.[9]

The Dog's Nose and the Olfactory System

The dog's nostrils have a complex structure and many important functions. In addition to being an organ for the sense of smell, the nostrils also aid with tempering, filtering, and humidifying the air that is inhaled and passes down into the lungs. The nostrils of dogs and human beings function for both breathing and sniffing. The dog's nostrils are remarkably well organized and far more advanced than our own.[10]

How odorants enter the nostrils and the structure of the nose itself, with its olfactory recess located farthest back in the nostril, are both important for dogs' keen sense of smell. When a dog inhales, the air is channeled along different paths. The rapid airstream (sniffing) travels to the olfactory epithelium (the olfactory mucous membrane), while the slower airstream (breath) travels to the lungs. A fold of tissue just inside the nose helps channel the two different airstreams.[11] When the dog breathes through its nose, the air passes through the respiratory region in the dog's long snout and subsequently directly into the lungs. When a dog sniffs, the air follows a side route, entering what we call the olfactory recess. The olfactory recess is covered by an olfactory epithelium containing genes for olfactory receptors (every single one of which is a protein produced by a specific gene), and olfactory receptor cells that absorb odorants.[12] Microsomatic mammals, such as humans and primates, have a different makeup, lacking this olfactory recess. The dog has agile nostrils that stretch when it is sniffing, and this movement opens an upper passageway that sends the air directly into the part of the olfactory recess farthest in the back. An enlarged olfactory recess very likely also increases the airstream for both inhalation and exhalation.[13] The air is filtered slowly forward through the sensory apparatus before it finds its way into the lungs.[14] Professor Gary S. Settles at Pennsylvania State University said that the entire system reminds him of the oil filter in a motor vehicle. The oil filter is located beside the engine, just like the olfactory recess. The oil moves directly to the part of the oil filter farthest back, and then slowly returns to the engine through the filter.[15]

The olfactory mucous membrane is spread across a labyrinth of bone structures called nasal conchae (turbinates) and is covered with millions of tiny hairs called cilia (or olfactory hairs). These are what capture odorants. When gaseous odorants come into contact with the olfactory membrane, they are dissolved in the layer of mucus. The odorants must be dissolved in water or fat in order to pass through the liquid in which the cilia lie, and the olfactory receptor cells receive the odorants in a dissolved state.[16] Odorants that are easily dissolved—such as dinitrotoluene (DNT), or dinitro—are released in the front part of the olfactory recess, while moderately soluble and insoluble odorants are distributed more evenly across the entire olfactory recess. How the odorants are deposited thus plays a part in compound recognition. Therefore, the dog's nose would not appear to be optimal for the detection of easily soluble substances (such as explosives) in that these are quickly absorbed upon entrance into the nostril.[17]

The composition of odorants determines whether they pass through the olfactory receptors in the nose, similar to how you need the right key to unlock the door to your house. The chemical formula and vibration pattern of odorants determine their scent. Some odorants can have very similar chemical formulas, but nonetheless have very different smells.[18] This can be understood as being comparable to our fingerprints or our DNA identity. After the odorants have passed the olfactory receptors, they are transformed into an electrical signal that travels via the olfactory nerve to the olfactory center of the brain, where the information is interpreted.[19] Not all substances can be identified by the nose: oxygen, nitrogen, and methane are odorless.[20]

The olfactory mucous membrane varies from one breed to the next, within each breed, and with age. The German shepherd has an olfactory mucous membrane area ranging from 96 cm^2 to 200 cm^2. A cocker spaniel has an olfactory mucous membrane area of 67 cm^2, and a fox terrier puppy can have an area as small as 11 cm^2.[21] The larger the surface area of the olfactory mucous membrane, the greater the potential for absorbing weak odor signals.

A dog has 872 functional olfactory receptor genes.[22] In comparison, humans have only 388.[23] If we compare with other animals, rats have 1,234 and mice have 913 functional olfactory receptor genes.[24] The kakapo, a rare bird found in New Zealand, has 667 working olfactory receptor genes.[25] In 2009 Stephanie Robin, a geneticist at the University of Rennes in France, and colleagues investigated the olfactory receptor genes of different dog breeds, and on the basis of this, they found that the Labrador retriever and German shepherd have a much greater potential as search and rescue dogs than the Pekingese and greyhound.[26] The number of olfactory receptor genes is also believed to influence the dog's ability to differentiate between very similar odors.[27] The number of pseudogenes (a gene without a function) can also be a determining factor for the capacity of a given dog breed's sense of smell. The more pseudogenes a dog has, the poorer its sense of smell. A boxer, for example, has 20 percent pseudogenes, while a poodle has 18 percent pseudogenes. We therefore assume that a boxer's sense of smell is inferior to that of a poodle.[28] In comparison, human beings have 67 percent olfactory receptor pseudogenes.[29]

The olfactory mucous membrane in the nose of a dog covers an area the size of the dog's skin surface, while in humans the surface area is the size of a postage stamp. The bloodhound is the dog breed with the most olfactory receptor cells—300 million![30] German shepherds have 220 million, the fox terrier, 147 million, and the dachshund, 125 million olfactory receptor cells.[31] The dog can detect odorants in far lower concentrations (the amount of a substance in a given volume of a solution or compound) than we can. They can smell some compounds with concentrations as low as one part per quintillion (1 in 10^{18}), which is much lower than the amount established for human beings.[32] In order to get a better sense of what one part per quintillion means, imagine that it is the same as 3 seconds in 100,000 years.[33] We can also illustrate this with another example. A gram of butyric acid contains 7 × 10 molecules. If the molecules are distributed evenly across all the rooms of a ten-story office building, we will only be able to smell the substance in one of the rooms. If the same gram of this substance filled the airspace over the entire city of Hamburg, a dog on the ground could detect it at a height of almost 92 meters.[34]

The Dog's Sense of Smell Compared to That of Human Beings

The dog has a much better sense of smell than human beings. In general, the dog's nose is 100,000 to 1 million times more sensitive than the human nose, while the bloodhound, which has the best nose, has a nose that is 10 to 100 million times more sensitive than ours.[35] The dog's rhinencephalon (smell-brain) is almost seven times larger than the human being's. Further, it has been proven that

- Thirty-three percent of the dog's brain interprets odors. Only 5 percent of the human brain interprets odors.[36]
- Including pseudogenes, dogs have a total of 1,094 olfactory receptor genes, and humans have a total of 802 olfactory receptor genes.[37]
- Adult dogs have an olfactory mucous membrane measuring 67–200 cm^2, while the olfactory mucous membrane of humans is only 3–10 cm^2.[38]
- Dogs can have 125–300 million olfactory cells. Humans have 5 million olfactory cells.[39]
- Dogs have 100–150 olfactory hairs per olfactory cell. Humans have 6–8 olfactory hairs per olfactory cell.[40]
- Dogs can smell some compounds at concentrations as low as one part per quintillion (1 in 10^{18}). For humans, the lowest concentration found is one part per billion (1 in 10^{9}).[41]

Factors That Can Have an Impact on a Dog's Odor Detection Outcome

There are many factors that can have an impact on a dog's odor detection outcome. It can be a matter of unintentional cues from the owner, what the dog has eaten, the amount of sleep it has had, its overall health, how the dog responds to us, and whether it likes to play and receive a reward. An uncomfortable or stressful environment can also influence the dog's performance. And, of course, the individual dog is also important. There are large individual differences within each respective breed.[42] Dogs can also have bad days, just like us. Young dogs seem to have a greater learning capacity. The older the dog is, the more diminished its ability to perform and learn will be.[43] In 2014 PhD candidate Lisa Wallis at the University of Veterinary Medicine in Vienna led a study which found that dogs in puberty (one to two years old) had a large potential for learning and training. Dogs at this age were thus found to be comparable to human teenagers: they learn quickly and effectively as long as one can get and keep their attention, something that is not always easy to do.[44] There can also be differences between the sexes when it comes to the sense of smell.[45] Female dogs have a better sense of smell than males, but this is diminished when a female dog is in heat.[46] The curiosity of female dogs also increases in correlation with training.[47]

The dog's sense of smell will be debilitated by different illnesses such as distemper and parainfluenza (kennel cough).[48] It is almost universally assumed that the canine nasal mite, a troublesome parasite, will reduce the dog's sense of smell, because it causes irritation in the sinus cavities and nasal cavity, and thereby inflammation and secretion of fluid/mucus.[49] This is comparable to the effect of a cold on our sense of smell. With nasal mite infections common and on the rise in many places, many dog owners give their hunting dogs a treatment against nasal mites before hunting season starts.[50]

If dogs eat less proteins and more fat, their sense of smell is improved. Dogs on diets containing a lot of corn oils were able to detect, for example, ammonia nitrate and 2,4,6-trinitrotoluene (TNT) more easily. This is probably due to the fact that the fatty substances enhance the functioning of the olfactory receptors. Another reason could be that they reduce the dog's body temperature, which in turn reduces panting and thereby improves sniffing. If nanoparticles of zinc are sprayed into the air, this can also help dogs to detect extremely faint odorants. Magnetic resonance imaging (MRI) has shown that there is an increase in activity in parts of the brain when these zinc particles are present.[51]

The odor composition and the concentration of the substance that the dog is searching for will also influence its odor detection results.[52]

Like us, dogs can suffer olfactory fatigue if they are exposed to the same odor for a long period of time or frequently.[53] Weather conditions such as the temperature, humidity, and wind velocity, as well as landscape topography and vegetation density also influence a dog's odor detection outcome (discussed in further detail in chapter 6, "On the Hunt").

The Dog's Advanced Nostrils

In 2010 professor of mechanics Brent A. Craven and his colleagues at Pennsylvania State University discovered that the dog's fantastic sense of smell can also be explained by the fact that dogs don't exhale when they are trying to sniff. This enables the dog to sniff faint odors without disturbing or destroying them. By using Schlieren photography, a special technique that registers how gases refract light at different temperatures, images can be made (up to 1,000 per second) that show the airstream produced by the nose of the dog.[54] Unlike humans, dogs can move their two nostrils independently. When a dog inhales, the air close to the nostril is drawn in, and the dog knows which nostril the air enters. The dog's nostril is more sophisticated than a pair of simple openings. Dogs have a wing-like flap in each nostril that opens for and shuts off the airstream moving through the nose. This flap determines the direction of the airstream in and out of the nose. When the dog inhales, there is an opening above and beside this flap. When the dog exhales, this opening closes and the air comes out below and beside this flap through another opening, enabling the dog to increase its collection of further odors. As a result, the warm air that is exhaled flows backward and away from the odor being sniffed and prevents the odor from being mixed into the air being breathed out. Because the air is warm, odorants are heated up and more easily converted into gas form, thereby reinforcing the gathering of odors. By keeping its nose close to the ground and sniffing in quickly, a dog can blow the heavier, non-volatile odorants up from the ground, bringing the odorants up into the air and into its nose.[55]

In 1996 biology professor Johan B. Steen at the University of Oslo and his colleagues found that dogs can sniff up to 210 times in the course of a minute while they are hunting, and can maintain the sniffing of odorants in the air for up to 40 seconds.[56] When a dog is searching for human odors, it sniffs six times a second, and the more difficult the task is, the more rapidly it will sniff.[57] A dog inhales approximately 60 milliliters of air per second through its nose, and if it sniffs six times per second, this means that it gathers 360 milliliters of air every second.[58] Dogs also sniff more when they are searching in darkness.[59] We can easily hear how a dog's sniffing speed increases. Just try listening when you give a dog an odor with which it is not very familiar. Although it may not always appear so, for a dog, sniffing is an active process. A dog can either sniff or pant, so a dog that is in good physical condition will be better able to find what it is looking for than a dog in poor shape.[60] This alone is incitement for dog owners to get out and train their dogs. Increased panting leads to reduced sniffing speed and therefore diminished olfactory abilities.[61] High temperatures will cause a dog to pant more, and it will become tired more quickly. It is important to be aware that different dogs have different levels of tolerance for heat, and for that reason they will pant at different degrees of intensity under the same environmental conditions.[62] If the conditions are dry, this can lead to dehydration and a dry nose, which will also impair the dog's sense of smell.[63] Proximity to livestock and exhaust and gasoline from motor vehicles can also influence the dog. Some dog trainers hold that the smell of gasoline can block the sense of smell for several minutes.[64] They maintain that if the dog has been riding in a car, it should be given 20 to 30 minutes to clear its nose before the search begins.[65] Other dog trainers have not experienced any problems with this.

When a dog sniffs, small bursts of air are blown out of the nose and sucked in again immediately. The air that is expelled is damp and can capture odorants on the outside of the nose. The same air is subsequently sucked in again.[66] When the dog exhales, a whirlpool is created that guides new odorants into the nose. This enables the dog to sniff more or less constantly. When human beings breathe out through the nose, we send the air out the same way it came in.[67] By expanding their nostrils, dogs can direct the airstream and send more information to the olfactory mucous membrane. The flow of air through the nose is much greater with normal sniffing than in a state of rest.[68] Rapid sniffing produces an airstream along the ground and propels the odorants up into the air and into the nose. When the dog sniffs more quickly, the volume of the airstream in the nostril also increases.[69] The nasal airstream is accordingly important for the perception of odors.

Both in humans and in dogs, the brain is divided into two hemispheres. Some of us are left-handed, others right-handed. Usually one of our eyes is dominant. Research led by Marcello Siniscalchi in the Department of Veterinary Medicine at the University of Bari Aldo Moro in 2011 showed that dogs use their nostrils differently according to the nature of the scent. When dogs sniffed at unfamiliar smells that were not dangerous (food, lemon, vaginal secretions from female dogs in heat, and cotton swab odors), first they used the right nostril and then switched to the left nostril to sniff at the odors again. This indicates that the right side of the brain was used when they sniffed at an unfamiliar smell. Once they had become familiar with the smell, the left side of the brain took over. When they sniffed sweat odors from veterinarians who worked at a kennel (in other words, stress odors), they used only the right nostril. In short, the left and right sides of the brain take in different kinds of information. The right side of the brain is associated with intense feelings, such as aggression, flight behavior, and fear (unlike other organ senses, olfactory pathways ascend ipsilaterally in mammals: odors in the right nostril is interpreted in the right hemisphere).[70] For most dogs, a veterinarian is a frightening person.

Human Beings' Sense of Smell Is Better than Formerly Believed

When somebody lights a cigar on the far side of the room, it takes a minute for the smell of the cigar to travel with the airstream to our nose and inside the nose to our olfactory receptors. You've already seen a person lighting the cigar and possibly also heard the sound of the lighter. Light moves at a speed of 300 million meters per second, sound moves through the air at 343 meters per second, and the airstream transports the odor of cigar smoke in a room at a speed of approximately one meter per second.[71] There is a 30 percent difference between your sense of smell and that of any random person with whom you might compare yourself. Each of us has a unique combination of olfactory receptors.[72] Because of this, we perceive smells differently. Some people have an oversensitive sense of smell, which is called hyperosmia, while others have a partially diminished sense of smell, hyposmia.[73] Not everyone has a sense of smell (anosmia); in fact, as many as 2 million Americans suffer from this. Research shows that this leads to diminished quality of life and greater risk of depression.[74]

However, most of us have a superb sense of smell; the problem is that we don't trust our nose. People can detect concentrations as low as 0.2 parts per billion (10^9).[75] This means that we can detect the odor of three drops of the odorant ethyl mercaptan—a substance that is often added to propane and butane and smells like boiled cabbage—in an Olympic-size swimming pool. If there are two such swimming pools, using our sense of smell we can determine which pool contains three drops of the odorant.[76] People can actually detect more than 1 trillion (1.72×10^{12} different odorants—in other words, a thousand billion, or a million million) different odors,[77] so it's a myth that we have a poor sense of smell. Human beings' sense of smell is in fact not very different from the olfactory system found in goats and guinea pigs.[78] Some of the things that humans can do include the following:

- Identify different dogs by their odors and recognize the odor of our own dog.[79]
- Detect the smell of fear in sweat.[80]
- Recognize our own scent among the scents of others.[81]
- Recognize the scent of our babies if we are mothers; and babies can recognize the scent of their mothers.[82]
- Recognize the scent of our children, siblings, family members, and our close friends.[83]

- Choose partners who have an advantageous genetic makeup by using our sense of smell.[84] Women who take birth control pills prefer the scent of men who have the same major histocompatibility complex (MHC) genes, while those who don't take the pill, prefer the scent of men who do not have the same MHC genes.[85]
- If we are men, tell by their scent whether or not women are menstruating.[86]
- Learn to identify new odors while asleep,[87] and we can become even more adept through training and practice.[88] Just think of how skilled perfume experts and wine tasters become after years of practice.[89]

The Dog's Nose and the Vomeronasal System

Dogs also have another olfactory organ, in addition to the olfactory system, which is called the vomeronasal organ, or Jacobson's organ.[90] The Danish surgeon Ludvig L. Jacobson described this organ in 1811 and is considered the person responsible for discovering it, although it was in fact the anatomist Frederik Ruysch from the Netherlands who discovered the organ first.[91] This organ consists of a pair of pouches filled with fluid, located in the roof of the dog's mouth, behind the upper incisors. In mammals, it is surrounded by bone or a cartilage capsule. The organ is thereby separated from the airstream that passes through the nostrils during normal breathing. The pouches filled with fluid also have olfactory receptor cells. There are separate neural pathways running from this organ to the brain. The function of this organ is not fully understood, but it probably plays an important part in the perception of pheromones (odorants secreted and detected by individuals of the same species).[92] The vomeronasal organ is believed to be particularly important in connection with reproduction in animals and recognition of kin.[93]

Many animals do something that is known as the flehmen response—they "curl" their lip. If you have a cat in the house and put the smell of urine from another cat on a cotton ball and hold this in front of the cat's nose, you will see how it "curls" its lip. This response helps transport the odorants to the vomeronasal organ. Dogs have been observed making rapid licking movements with the tip of the tongue and moving the tongue repeatedly up against the roof of the mouth where we find the canal opening to the organ. When a dog does this, it keeps the lips partially curled (the flehmen response) and holds the incisors of the upper and lower jaw slightly apart. This activity makes it easier for the dog to transport the heavier, non-volatile odorants to this organ.[94] The teeth of male dogs click and chatter when they come into contact with female dogs in heat. If the females have deposited a scent, the male dog will also lick this, very likely to transport the odorants to the vomeronasal organ.[95] It is the substance methyl-p-hydroxybenzoate from the vaginal secretion of female dogs in heat that causes male dogs to try to mount females.[96] A related substance is used as a preservative in cosmetics, shampoo, and hand lotions, and anyone who uses any of these products may find themselves subjected to the amorous attentions of interested male dogs.[97]

In 2012 research scientist Daisy Berthoud at Anglia Ruskin University in England observed this flehmen response behavior more often in dogs that were not neutered than in neutered dogs when they sniffed at urine markings from other neutered and unneutered male dogs in a boarding kennel. When the dog was at home in its own backyard, it was mostly unneutered male dogs that exhibited flehmen response behavior when they sniffed urine from an unneutered and a neutered male dog. It is therefore probable that neutered dogs do not receive complete information when they sniff at an odor marking from another dog.[98]

The Dog's Olfactory Memory

Scientists have proven that there is a close connection between odors and memory. Everyone has certainly experienced how a specific smell can stir up memories of places, people, or events. For most of us, odorants remind us of something from the past, and especially things associated with strong feelings.[99] That is how it is for dogs too: they have a brilliant memory for odors. Following repeated training with the same odor, a dog will be able to recognize it. They can easily learn to react to at least ten different odors[100] and will remember odorants for a long time. A dog trained to find feces from the San Joaquin kit fox continued to react to this odor 671 days after the most recent exposure, in other words, almost two years later. Dogs that are systematically trained with certain odors develop more olfactory receptors for these odors and thereby increase their sensitivity.[101]

3

A Good Judge of Character

On October 3, 1992, in Upper Merion Township, Pennsylvania, dog trainer and animal behaviorist consultant Susan Bulanda, who has written several books about dogs, was asked to help out in the search for a missing man. The search had already been under way for two days. A search team of hundreds of people had been combing the area where the man had disappeared three days before. The police had found his car in the parking lot of a railway station. In the parking lot, they also found a bloodied shirt that was confirmed to belong to the missing man. They put the shirt in a paper bag and left it in the location where it had been found. There was a large forest near the station, where the man had told his wife he would be bow hunting for deer. The search crew—with and without dogs—had searched the forest without success. Bulanda was summoned because she had the best dog in the region. When Bulanda arrived on the scene, she checked the shirt in the paper bag and saw that the bloodstain was the same size and shape on the front as on the back, but there were no holes in the shirt. Bulanda gathered odors from the untouched clothing inside the car on a cotton ball. These clothes belonged to the missing person. Her dog Scout, a Beauceron, is trained to track human beings, to find cadavers on land and as a disaster search dog. He was allowed to sniff the cotton ball, and right away he began to move away from the car and toward the paper bag. He sniffed at the bloodied shirt in the paper bag for one second and subsequently went directly toward a path in the forest on the other side of the road. Scout followed the path and eventually reached a fork, upon which he followed the path leading right. Later there was another fork in the path; this time he went to the left. Scout led Bulanda up along a river, farther and farther into the forest. After a while, he stopped and sniffed the air before making a sharp turn to the left. Finally, they reached another parking lot, where Scout stopped and indicated that the trail had come to an end. In Bulanda's report to the police, she concluded that the missing man had driven away from this location in another car. Three days later, the man turned up in another country with a female companion. It turned out that the missing man had staged his own death so he could run away with his new girlfriend.[1]

Tracking Human Beings

Dogs were used to track down runaway slaves in the Far East as far back in time as 1000 BC.[2] Christopher Columbus used dogs to find and kill native Indians when he reached America in 1492, and the Spanish used dogs in South America to track down runaway plantation workers.[3] Since 2012 dogs have been used to track down poachers in the Virunga National Park in the Congo in Africa.[4] To better understand how dogs manage to follow a human trail or find a human being by sniffing airborne scents, it is important to understand how odorants are emitted and the sources of odor that dogs use in the context of different work tasks. There is, in fact, a large amount of information found in the odors we spread around us.

In 1936 the scientists Konrad Most, director of the Canine Research Department of the German army high command, and Gustav H. Brückner at the University of Rostock built a chair lift and a tracking wheel in a forest in Germany. The tracking wheel was two meters in diameter, and the scientists wanted to separate the different components in a trail they presumed the dogs could use. A clog was attached to the wheel, which thereby created a footprint at regular intervals. The wheel created a trail of crushed plants and disturbed soil without leaving behind a human scent. On the other hand, if a person sat in the chairlift with his or her feet up, no footprints would be made and thereby no scents would be emitted from the earth or vegetation. The person who was seated in the chairlift, however, emitted an odor in the form of dead skin cells.[5] When dogs are following the trail of a human being, they can use both the scent from this person and the scent from the destruction of vegetation or disturbed soil. If a patch of earth is moved or disturbed, the scent will be altered. If one sniffs grass that is relatively undisturbed, treads on it, then smells it again, the scent will be different; dogs pick up on this much better than we do.[6]

Dogs recognize different human beings through the scent of skin cell flakes, which are microscopic particles that human beings shed constantly. Unbelievable as it may seem, we shed approximately 40,000 dead skin cell flakes a minute,[7] depending upon the individual's level of activity and their emotional and physical condition.[8] Formerly, scientists believed that skin cell flakes fell directly off the human body, but later discovered that there is an airstream close to the surface of the skin. The airstream provides a transport system for skin cell flakes and bacteria. It starts at the feet, moves upward along the body, and finally takes off at the top of the head.[9] The airstream around a person has an impact on how the skin cells swirl and fall off. If we stand with our legs apart, this affects the structure of the airstream behind us. Swinging your arms, on the other hand, has little impact on the structure of the airstream. When we are walking, we have two separate airstream regions behind us—one behind the back and another behind the legs—and there is a considerable downdraft behind our backs that spreads the lower part of the airstream.[10] The skin cell flakes can be transported a distance of at least eight meters away from the body.[11] The distance away from a trail that odorants will be transported depends upon the particle weight (molecule weight) and wind.

The activity of the microorganisms (bacteria, fungus, and parasites) on the skin surface contributes to the creation of odorants. A skin cell flake is made up of one or more dead cells, roughly four bacteria, and some bodily secretions.[12] The composition of microorganisms varies considerably in humans and is what gives us different odors.[13] The respiratory system also emits odorants through the breath. We all know about bad breath, caused by bacterial activity in our mouths. The smells from the environment also contribute to changing our odor. Our odor can be affected by our diet, and if we eat meat, we smell worse.[14] If we smoke, take medication, use different perfumes, deodorants, and soaps/shampoos, this can also change our odor.[15] The people with whom we have daily contact (such as our spouse and children) can also affect the way we smell, because they transfer bacteria onto us. We can also pick up bacteria from our contact with food, water, air, and other objects in our environment. Even our dog can contribute bacteria. Adult dog owners have several of the same microorganisms as their own dogs on their skin.[16] Pregnant women who live with dogs are twice as likely to have intestinal bacteria in their vagina as other women.[17] It is of interest to note that pregnant women who have dogs in their home are less at risk of having children with allergies.[18] Exposure to many different microorganisms can be good for us.

The Skin Glands Produce a Scent

The skin is our largest organ and represents 12–15 percent of our body weight.[19] There are first and foremost three types of glands that produce odorants from our skin.[20] The apocrine sweat glands are the most important and produce different lipids, proteins, and steroids. They are found predominantly under the arms, on the chest, and around our genitals. The eccrine glands are the true sweat glands. They are found all over the body and are especially concentrated on the palms of the hands, the soles of the feet, and the forehead. The sebaceous glands (or holocrine glands), which produce oily secretions (such as fatty acids and wax esters), are found all over the body but predominantly where we have hair, such as on the head and face. Most of us have experienced having somewhat greasy hair or pimples, and it is these holocrine glands that produce excess secretion. This increase in the production of secretions in humans can be due to hormones, pregnancy, menopause, climatic conditions (warmth or humidity), pollution, and/or stress.[21] It is well known that the head has a characteristic odor that can vary in its intensity from one person to the next.[22]

Human beings have approximately 5 million glands that produce the secretion that is transported to the surface of the skin through the pores.[23] At birth, our sweat is virtually odorless, since it is mostly water, but the microorganisms will eventually produce a scent. When our sweat smells bad, it is usually from our armpits and feet, and is transferred to the clothing and shoes we wear.[24] The odor of our sweat is therefore an important source for dogs that are searching for and tracking us.

In 1990 the French scientist Jean-Claude Filiâtre and his colleagues at Franche-Comté University demonstrated that when a dog sniffed a doll wearing the clothing of a child it did not know, the dog sniffed more in the genital region. When the dog was presented with a doll wearing the clothing of a child it did know, the dog sniffed in many different areas. This means that a dog will first identify the child by sniffing at the genital region, and then attempt to establish the child's emotional and mental state by sniffing the body and then the head.[25] In another study done by the same scientists the following year, they found that a dog will always sniff a familiar person more than an unfamiliar person. When a dog met its adult owners, it sniffed at many different body areas. In meeting with an unfamiliar person, again the dog sniffed the genital area first to determine the identity, and then sniffed mainly at the head and hands to discover something about the person's emotional state.[26]

The different parts of the human body have different smells because the number and types of skin glands vary. Human beings also have different types and amounts of microorganisms, and we do not have the same access to oxygen everywhere on our body.[27] If a dog is allowed to sniff a T-shirt, it will not necessarily recognize a pair of jeans from the same person.[28] The volatile odorants from the hands, the saliva, breath, blood, and urine of the same person are too dissimilar to be useful in comparisons, such as in a scent lineup (see chapter 7, "Police Work").[29] The volatile substances from the skin on the upper part of the back and the forearms are quite similar, but there are also significant differences. Both of these body regions have sebaceous glands, but these glands are much more concentrated on the upper part of the back than on the lower part of the arm.[30] If a person puts on underwear that has been worn first by somebody else, and then gets into a car belonging to a third person, it is said that a dog will be able to differentiate between the scents of all three people if it is given a scent sample from the car seat.[31]

We emit quantities of volatile substances from our feet, which make it possible for a dog to follow our trail.[32] People who have a strong foot odor have many fatty acids, and the substance isovaleric acid is only found in people with foot odor.[33] These volatile fatty acids pass through the soles of shoes and are deposited on the ground. Foot sweat also passes through rubber boots and is deposited on a trail. The front part of the foot will give off the most odor because the greatest number of sweat glands are accumulated there. The oily secretion between the toes is more prone to attack by bacteria. The amount of odorants emitted depends on whether or not you are wearing socks with your shoes and the kind of materials your shoes are made of.[34]

In order for dogs to find an object, odorants from the skin glands must have been deposited by the person who has touched the object. Odorants from the hands come from the eccrine and sebaceous glands.[35] Research scientists Aline Girod and Celine Weyermann at the University of Lausanne in Switzerland found in 2014 that fingerprints from twenty-five people contained 104 different fatty substances. Some people emit many substances, while others emit only a few.[36] If we hold a stick in our hand for just one or two seconds, this is sufficient to enable a dog to identify it. Contact with a single fingertip is enough. The dog will also manage to determine whether a stick lying on the ground has an odor on the sides, top, or underneath.[37] Dogs are able to identify our scent on fragments of glass that have been stored indoors for a month and outdoors for two weeks.[38] Even if a metal pipe carrying a human scent has been exposed to temperatures of up to

TABLE 1: The most important sources of odor in humans and predominant use

Our most important sources of odor	Used predominantly by
Skin cell flakes	Dogs searching for humans
Sweat and sebaceous glands	Dogs searching for humans
Breath	Rescue and medical detection dogs
Saliva	Police dogs
Blood	Rescue, police, and medical detection dogs
Semen	Police dogs
Urine	Rescue and medical detection dogs
Feces	Medical detection and environmental dogs

Sources: B. Schaal and R. H. Porter, "Microsmatic Humans Revisited: The Generation and Perception of Chemical Signals," *Advances in the Study of Behavior* 20 (1991): 135–99; M. Shirasu and K. Touhara, "The Scent of Disease: Volatile Organic Compounds of the Human Body Related to Disease and Disorder," *Journal of Biochemistry* 150 (2011): 257–66, doi:10.1093/jb/mvr090; M. Kusano, E. Mendez, and K. G. Furton, "Comparison of the Volatile Organic Compounds from Different Biological Specimens for Profiling Potential," Journal of Forensic Sciences 58 (2013): 29–39, doi:10.1111/j.1556-4029.2012.02215.x.

800–900°C, dogs will be able to identify our scent. In one experiment, however, none of the dogs succeeded in identifying the scent when the temperature was 1000°C.[39] In 2014 the scientist Petra Vyplelová at the University of Prague discovered that it is not necessary for us to be in contact with an object for our odorants to be deposited. If we hold our hand five centimeters above a rag for three minutes, a dog will be able to identify us by sniffing the rag. In other words, our scent is deposited in the environment around us even if we are not in direct physical contact with our surroundings.[40]

Professor of behavioral biology Tadeusz Jezierski and his colleagues in Jastrzębie-Zdrój, Poland, demonstrated that it is easier for dogs to identify smells from women's hands than those from men's hands. This may be due to the fact that it is easier for dogs to recognize the scent of women or that dogs are more attracted to the scent of women.[41] This is supported by a chemical study that established that there were more chemical compounds found on women's hands than on men's (58 versus 46 compounds), and some compounds were only found in the one gender.[42] A dog trained to detect progesterone preferred the smell of pregnant women over the smell of men or women who were not pregnant.[43]

Human Odors Survive Bomb Explosions

In 2010 scientist Allison M. Curran and her colleagues at Florida International University wanted to find out whether human odors survive the extreme conditions of an explosion.[44] They collected odors in a car from the steering wheel, a door, and a nylon bag after a car bomb had exploded. Two people had been in the car—a terrorist and his driver. It was therefore possible that odors from both people had been deposited on these objects. After having activated the bomb, the two men walked to a city a few kilometers away. There they each went their separate ways and walked to different buildings. Six other people were also in the area. The dogs did not know any of the people. The dogs' task was to track down the terrorist and the driver four hours after the explosion. First, they were allowed to sniff the odor that had been collected. When the trail divided and went in two directions, they were supposed to choose one of them. The wind direction on this day was favorable, coming from the driver and toward the dogs. As a result, all twelve of the equipages (dog and handler) followed the trail of the driver. Eight of these managed to locate the driver at the end of the trail. In another similar trial using improvised explosives, all except one of the equipages were able to follow the trail and locate the terrorist or the driver. In these two trials, the dogs managed to find the terrorist or driver in 82 percent of the trials.

The Direction of a Human Trail

For most carnivores, it is important to be able to locate and track prey in the right direction in order to survive. But how do dogs know how to determine the direction on a trail? Eight hundred years ago, Icelandic author Snorre Sturlason described how two Norwegians, who were being held prisoner by the Swedes in 1026, escaped from prison and tricked their pursuers by fastening reindeer hoofs back-to-front beneath their shoes. According to the story, the Swedish tracker dogs followed the trail in the wrong direction and ended up near the place where the two Norwegians had been imprisoned. Here they found a big hole in the fence. Subsequent research carried out by biology professor Johan B. Steen at the University of Oslo and scientist Erik Wilsson at the Swedish dog training center in Sollefteå, however, has shown that dogs are not fooled by a trail that looks like it is moving backward. In another experiment, the dogs were unable to determine the direction of a trail laid on grass or asphalt. This was because the trail was created either by dragging a pair of shoes along the ground or by walking with such small steps that the heel hit the ground in the same place where the tips of the toes from the previous step had been.[45]

In 2005 Professor Peter Hepper and research scientist Deborah L. Wells at Queens University in Belfast showed that when the scent of footsteps is removed, dogs are not able to determine the direction of the trail.[46] It turns out that dogs use the individual scent deposited in footprints to determine the direction. In a normal footstep, the heel is set down before the toe, but this was not what the dogs used to figure out the direction. Dogs can "read" when a footprint is created and assess the time the footprint is made for the different steps, and thereby determine the direction of the trail. A footprint needs only one to two seconds to change (depending upon the environment), and because of this the difference in the odor of each footprint is sufficient to enable the dog to determine the direction of the trail.[47] Experienced dogs are extremely certain when they determine the direction of a trail, and experiments have shown that dogs can determine the direction of a human trail on the basis of only two to five consecutive footprints. They manage this on both a grass surface and a hard surface such as asphalt.[48] It may also be that the fresher the footprint, the stronger its odor, so the dog follows the trail in the direction where the scent is the strongest. When research scientists Steen and Wilson walked at one step per second and their footsteps were separated, the dogs managed to determine the direction of the trail. The scent of each footprint is then, theoretically speaking, 1/1,800 different from the preceding one and the dogs can detect this difference.[49] An alternative explanation is that more recent footsteps have a weaker scent than older steps. This can be due to older footprints having been affected by the environment and bacteria and thereby "putrefied." Newer footsteps therefore smell less putrefied than footsteps created earlier, which enables the dog to determine the direction of the trail by moving in the direction of the least putrefied scent.[50]

Individual dogs differ greatly when it comes to their abilities for determining the direction of a trail.[51] In another study carried out by Wells and Hepper in 2003, relatively few of the dogs managed this (8 out of 22), and their abilities proved to be age- and sex-dependent.[52] The male dogs found the correct direction of the trail more often than the females, and the younger animals were more adept than the older ones. Male dogs are probably better because they are usually more involved in odor-based activities such as tracking females and scent-marking their territory.[53]

Human beings' sense of smell deteriorates as we grow older and becomes less sensitive to heavy substances.[54] This would also appear to be the case for dogs.[55] The olfactory mucous membrane becomes thinner, and the number of olfactory receptor cells decreases when dogs age beyond fourteen years.[56] The dogs' performance outcomes also depended upon the direction in which the trail was made. The dogs were better at determining the direction of a trail made from left to right, than one made from right to left.[57] This can be due to wind conditions or the dogs' and handlers' preference for the direction to the right.[58] Dogs use the right nostril when they are sniffing unfamiliar odors, so presumably they more frequently followed the trail to the right.

Human Beings Can Also Follow a Trail

In a study done at the University of California, Berkeley, in 2007, PhD candidate in biophysics Jessica Porter and her colleagues laid a 10-meter trail of chocolate in a grass field. They then asked students to follow the trail using only their sense of smell.[59] The students were blindfolded, wore heavy gloves on their hands, and had earplugs. Most of the students were able to follow the trail. They tracked in a zigzag pattern similar to the way dogs do. When the students used only one nostril, they were slower and not as accurate while students who used both nostrils were quicker and more accurate. The faster they tracked, the faster they sniffed to acquire the same information, and the more they practiced, the more they improved. These experiments show that the use of both nostrils is important in following a trail. In other words, stereo-sniffing is important for discovering the source of an odor, in the same way that a pair of ears is important to locate the source of sound and two eyes are important for depth vision.[60]

Jayne M. Gardiner, PhD candidate in sensor biology at the University of Southern Florida, and Jelle Atema, professor of biology at Boston University, carried out odor trials on sharks in 2010. They found that the sharks did not compare the odor concentrations from both nostrils to navigate, as we have long believed, but instead registered when the odor arrived in the nostril. Sharks can detect small delays in the time when odorants meet the one nostril in relation to the other, and thereby turn in the direction of the side that picks up the odor first. These results contradict the common conception that an animal follows a trail based on differences in the concentration of odorants.[61] Like sharks, dogs also know which nostril the odor enters.[62]

What Kind of Information Is Found in Our Odor?

As far back as 1887, the English evolutionary biologist George J. Romanes documented that dogs were able to differentiate between people based on their odor.[63] The odor we produce reflects our internal physiological and metabolic status in a complex interaction between our genes and the environment.[64] It is not just the hands of individual people that are different, but also the saliva, breath, blood, and urine.[65] Many studies have shown that the major histocompatibility complex (MHC) plays an important role in the recognition of individuals and family members.[66] It also plays an important part in the attractiveness of human beings.[67] All of us—with the exception of identical twins—have different genes, and MHC variation can therefore provide dogs with information about a person's identity.[68]

Our odor is also dependent upon our sex and race.[69] Men have more and larger apocrine sweat glands than women, and the androsterone level in the armpits is therefore also much higher.[70] Each of the sexes also has its own particular odor because the bacteria found in the armpits are different.[71] It is the armpits in particular that give each of us our own distinctive odor, since all three types of skin glands are found here in large quantities. Many of the odorants here are also in our urine and saliva.[72] Asian people have fewer and less active apocrine sweat glands and therefore less of an odor. They also have another type of earwax.[73] Because of genetic differences, the armpits of Caucasians and Africans have a stronger odor than those of Asians. Our body odor also changes with age, especially during the period from childhood until puberty and from midlife (around the age of 39) until we reach old age.[74] For this reason, sometimes dogs won't give indications for the odors of children and teenagers if they have been trained to find adults. It is therefore important to train dogs for different age groups, for children, teenagers, and adults.

Women's odor changes during menstruation and pregnancy. Our emotional state and the state of our health also have an effect on our odor. Illnesses often produce a characteristic odor, such as the smell of cancer (see chapter 10, "Medical Detection").[75] The human odor is also intensified by extreme happiness or sadness.[76] Many people even maintain that dogs can smell the difference between tears of joy and the types of tears that are secreted due to sadness or fear. This is supported by the 2011 research findings of Shani Gelstein and colleagues from the Edith Wolfson Medical Center in Israel, which showed that men were unconsciously affected by the smell of the tears of women who had seen a sad movie. The men's sex drive was reduced (their testosterone level went down) by the women who were shedding tears.[77] And in that people can react to the smell of tears, there is every reason to assume that dogs will do the same.

The odor of human beings changes in response to states such as stress or fear, and dogs perceive this. If a person is feeling stressed, the substance cortisol is secreted. Police dogs can find the people they are searching for more easily than rescue dogs can because the former more frequently track criminals, who have a stronger odor. Criminals often sweat a great deal because they are afraid and are very physically active during a crime. Criminals who commit sex crimes are usually very aroused, which also produces a stronger odor. These individuals are therefore easier to track. Some people simply smell more than others, and the stronger their smell, the better it is for the dogs trying to find them. The odor emitted by individuals with Alzheimer's disease, who have a limited emotional capacity, is weaker than that of other people, which makes it much more difficult for dogs to find them. People with autism are also more difficult to track down.[78]

Our odor contains information about the following:

- individual identity[79]
- sex[80]
- race[81]
- age[82]
- kinship[83]
- reproductive status (pregnancy, ovulation, and menstruation)[84]
- illnesses[85]
- emotions such as happiness, sadness, stress, fear, and anxiety[86]
- psychiatric conditions[87]

Dog Brain Scans

A number of canine research scientists have now implemented functional magnetic resonance imaging (fMRI), a brain-imaging technique that depicts changes in activity in areas of the brain in both humans and animals. This method can provide helpful information about the connection between cognitive activity (which has to do with comprehension, understanding, and thought) and the location of the functions in the brain.[88] In 2012 professor of neurobiology Gregory S. Berns at Emory University in Atlanta trained thirteen dogs to lie completely still for an fMRI.[89] The dogs lying in the fMRI machine were allowed to sniff the urine of both unfamiliar and familiar dogs and from both unfamiliar and familiar people. They were also allowed to sniff their own urine, but no brain activity was registered in that case. The scientist found this comparable to our own breath, the smell of which we don't recognize. Unknown odors from people and other dogs required a lot of brain activity. Familiar odors from people or dogs, on the other hand, did not require much brain activity. When the dog sniffed urine from a person it knew, the part of the brain that stores memories was activated. This experiment showed that the dog managed to identify important people in its life even if the person in question was not physically present. The odor of the person was stored in the dog's brain.[90] This technique opens up an abundance of new possibilities for further study of the dog's brain.[91]

Can Dogs Tell Identical Twins Apart?

Many research scientists have carried out odor studies on identical twins to find out whether dogs can tell them apart.[92] Identical twins are genetically the same but have different fingerprints. But do they also have different odors? The studies have produced mixed results, probably because the dogs used had received different types of training. In 2011 research scientist Ludvik Pinc and his colleagues at the University of Prague carried out new trials in which the dogs were put through an intensive and more advanced type of training.[93] Ten police dogs were given the task of comparing the odor from a cotton ball that had been in contact with the abdominal region of identical twins and dizygotic twins, respectively, for twenty minutes. Both pairs of twins lived in the same environment and received the same food. In each trial, the odor of one of the twins was used as a starting point, and the dog was supposed to find out whether the odor from any of the glass jars on a scent detection board came from the other twin.

All of the dogs were able to differentiate the odors of all the dizygotic twins. Surprisingly, they were also able to differentiate the odors of the identical twins. All of the dogs were also able to find two odors that were collected from the same individuals. The results showed that the dogs were able to differentiate between individual odors from identical twins, even though they have lived in the same environment and eaten the same food, and this was also the case when the odor was not presented to them at the same time.[94] The level of training the dogs received was, in other words, of decisive significance to the results.

Tutta Sniffs Out Nervous Customers at the State Liquor Store

In September 2013, the Norwegian state liquor store, Vinmonopolet, launched a national campaign to discourage minors from using a fake ID to buy alcohol. In Norway the legal drinking age is eighteen years old for wine and twenty for liquor. Everyone under the age of twenty-five must present valid identification when they purchase alcohol, and the consequences of being caught using a fake ID are relatively severe, ranging from a fine to a black mark on one's record, to imprisonment. Vinmonopolet turns away approximately 10,000 people annually at the cash register because they are underage or because they don't have proper identification. In a study carried out by Vinmonopolet, one out of ten young people under the age of twenty-three reported that they had tampered with or borrowed an ID to buy alcohol. Vinmonopolet decided to address a serious problem in an untraditional and humorous way.

In the summer of 2013, the dog trainer Torun Thomassen trained the dog Tutta to detect the odor of fear emitted by persons carrying a fake ID. People in this situation may, of course, be nervous for other reasons, but if Tutta indicates detection of a target by freezing in place, the customer in question receives an informative brochure explaining the consequences of using a fake ID. Vinmonopolet thereby gives young people a friendly reminder, while simultaneously spotlighting the problem. Thomassen used samples containing both high- and low-stress levels from the same person and taught Tutta to ignore the samples with low values. Young people from a local athletics club had contributed the odor samples, and young people doing their compulsory national service at the armed forces' dog school had volunteered to help Tutta with the training. When a person is planning to do something illegal, such as use a fake or borrowed ID, the brain initiates a response to cope with the stressful situation. The substance cortisol is emitted by the adrenal glands into the blood and subsequently out through the skin, especially in those areas where sweat is usually secreted. Tutta was trained with this substance on both a training platform and scent detection board and subsequently on people in the age group 16–20 years. Tutta has traveled to different state liquor stores throughout Norway, such as in Bergen, Kristiansand, Oslo, Stavanger, Trondheim, and Tromsø. At the Vinmonopolet in Porsgrunn, Thomassen attached a cotton ball containing high cortisol levels to the back of the knee of a volunteer. On another person she attached a cotton ball with normal cortisol values. Tutta went directly to the person with the high cortisol values and indicated detection by freezing in place. Tutta was then rewarded with her well-deserved Kong rubber chew toy.[95]

4

Pet Finder

We don't always know where to start searching. Is the dog still alive, is it injured, or is it dead? Detective Kat Albrecht from Seattle was the first person in the world to begin doing professional searches for missing pets.[1] One winter day in 1997, her bloodhound AJ disappeared. He had dug a hole beneath the fence of his pen and run away. Perhaps he got a whiff of an irresistible odor that he just had to follow? Kat had lost not only a pet, but also her partner on the police force. She grew frightened and feared she would never see AJ again. She contacted a colleague who worked with Thalie, a golden retriever search and rescue dog. Thalie was allowed to sniff AJ's bed and was subsequently sent out to follow AJ's trail. Within fifteen minutes, Thalie had found AJ in the neighborhood. Then Kat understood that dogs not only understand the cue "Sniff this pillowcase and find the person," but they can also understand "Sniff this bed and find the dog that slept here" (see chapter 5, "Search and Rescue").

The Bond between the Dog and Its Owner

Since you are reading this book, it is highly likely that you are a "dog person." Dog people are said to have a personality that is different from that of cat people.[2] Dog people are known to be more outgoing, positive, and conscientious than cat people. Having a dog as a pet has psychological and physiological health benefits, and dog people know that a dog is a positive influence on our health and well-being.[3] Many dog owners go for walks in all kinds of weather, which means they exercise regularly. Dogs make us happy, keep us healthy, and give us an experience of being loved. When we are having a good time with our dog, this causes the levels of the hormone oxytocin to increase in both us and our dogs. Oxytocin has a tranquilizing effect, and the heartbeat, blood pressure, and stress level are all reduced. The likelihood of having a heart attack is substantially diminished, but should it nonetheless occur, the probability of survival is three to four times greater if we have close contact with a dog[4] (for more on this subject, see chapter 10, "Medical Detection").

The experience of being greeted by a happy dog wagging its tail when you come home after a long day at work is wonderful. The amount of time dogs are left at home alone, however, will influence how they greet us when we see them again. If we are away for two to four hours, the dog will wag its tail more, lick its lips more, and also wiggle its body more than if we are away for only thirty minutes. Also, the longer a dog has been alone, the more its heart rate will increase.[5] People who spend substantial amounts of time with their dogs will end up with dogs that seek a lot of contact when they are reunited with their owner. This can be due to the fact that the dogs are rewarded by the owner for seeking physical contact, or that dogs who are more attached to their owners are not as accustomed to being alone and therefore react with more contact-seeking behavior, the way an insecure child will do. Even if a person has a strong emotional tie to his or her dog, it unfortunately does not appear that dogs are equally attached to their owners.[6]

Searching for Pets

Dogs can also track down pets.[7] In general, we can say that dogs track animals in the same way that they track humans. When dogs run away, it can often be very difficult to locate them. Most people will do everything they can to find a missing four-legged friend. If they are accompanied by a trailing dog trained in scent discrimination, this job will be much simpler. A scent-discriminating trailing dog will be presented with an object the missing dog has been in contact with to sniff at before the search begins.

Dogs that enjoy playing with other dogs can be suitable pet finders. After her experience with a pet finder, Albrecht started her own company, Pet Hunters International. Albrecht and her dog AJ have helped more than 1,800 pet owners find their pets: besides dogs, they have located cats, horses, snakes, tortoises, ferrets, iguanas, and even a gecko that escaped from an aquarium. She has trained more than 200 pet detectives in the United States, Canada, Mexico, Japan, Italy, Australia, and Ireland, who have in turn helped thousands of pet owners recover their missing pets.[8]

Can Dogs Find Their Way Home over Large Distances?

There are a number of indications suggesting that dogs have an incredible aptitude for finding their way home again. A family in the United States brought their young dog with them on a train, after deciding to move from Canon, Colorado, to Denver. The train traveled over several large mountains and rivers during the 257-kilometer journey. The dog clearly did not like its new dwelling and wanted to go home. A week later it was back. A neighbor heard the dog's intense barks of joy and adopted it.[9] How had the dog found its way home? The findings of research scientist Vlastimil Hart and his colleagues at the University of Prague in 2013 indicate that dogs may have a "sixth sense" that enables them to "read" the earth's magnetic field. Like many other animals, dogs are sensitive to the earth's magnetic field. Dogs prefer to align their body along a north-south axis while defecating and mark with urine when the earth's magnetic field is calm (which it is for approximately 20 percent of the day).[10]

The Sources of a Dog's Odors and Marking Activity

To understand how dogs manage to find other dogs, it is important to know something about the sources of the dog's odors and its marking behavior.[11] Odor signals are produced by the dog's entire body, both inside and out. Dogs have far more sources of odor than most people are aware of. Dogs are constantly shedding dead skin cell flakes that carry the scent of their skin or that have been altered by bacteria. Odorants from the dog can also come from the breath, hair, genitals (urine and vaginal secretions or foreskin secretions), the rectum, feces, glands, or body fluids (e.g., blood and saliva).[12]

The dog can deposit odor markings both passively and actively. Passively, odor is deposited from the glands on the dog's paws as it walks. The dog can actively mark by urinating on a lamppost, a tree, or a building, by defecating, or by rolling on the ground. Anal sac secretion can also be sprayed and deposited on the ground or on feces. The dog will also deposit body odor on vegetation or other surfaces around its trail. A paw print causes the release of fluids from plants, and bacteria breaks down the dead plant cells on or in the soil.[13]

Rolling in Feces, Cadavers, and Other Foul-Smelling Materials

There are many myths about why dogs roll around in substances that have a strong odor, rubbing, in particular, their jowls, throat, and neck on the smelly substance. Are they doing this to (1) hide their own odor from potential predators, (2) inform group members of a good food source, (3) deposit their own odor, (4) perfume themselves with and revel in the odor, (5) ascend in rank, (6) change their own odor, or (7) create a diversion for group members so the recognition ceremony will be friendlier for those who come home, and so they forget everything else? Are they trying to defend their food, which can be a completely or partially rotten animal or cow dung, by marking it?[14] Why don't they mark these with urine or feces instead and then roll around beside it?[15] And why don't they just eat the dung, there and then? Unfortunately, as of 2014, no scientific studies have been done to explain this behavior of dogs. The first explanation, to hide their own odor from potential predators, is probably the most credible.[16]

Odors from Skin Glands

Dogs have sebaceous (or holocrine) glands that empty out into the hair follicles. These glands are found on the back of the neck, the back, and the tail. They are also found on the transition area between the skin and the mucous membrane on the lips (the corners of the mouth), in and around the ears, and on the labia and eyelids (tear glands).[17] These glands produce an oily secretion called sebum. Dogs also have apocrine sweat glands that are connected to the hair follicles and emit a water-like secretion.[18] These are found on the face and lips, and behind and between the toes, and become active during puberty at six to fourteen months old.[19]

The hairs on dogs' shoulders and back stand up when they are afraid or when it is cold, making them look larger than they actually are. This is a visual signal that is easy for us to recognize. Another dog perceives this as meaning that it would be a good idea to back off. When we "get the chills" from the sound of somebody singing beautifully, or when we are frightened by a scary scene in a horror movie, the hairs on our arms also stand up. We get goose bumps from such intensely emotional situations. The same thing happens to the dog's hair. When dogs get frightened, for example, an involuntary reaction occurs causing the release of hormones, such as adrenaline. These hormones cause the muscles to stretch the skin, so the hairs stand straight up and secretions from the sebaceous glands and the apocrine sweat glands are released.[20] It is therefore likely that these secretions say something about the dog's emotional state.

The eccrine glands are the true sweat glands and are not connected to the hair follicles. In dogs they are found only on the paws, but there are more of these glands than apocrine sweat glands.[21] The secretion is produced first and foremost through activity and heat stress, but can also be emitted involuntarily from the nervous system.[22] Dogs often scrape the ground with stiff legs when they discover a urine marking or feces. It is very likely that dogs are depositing a new scent marking from the apocrine sweat glands between the toes and from the true sweat glands on the paws. In addition to the odor from the glands on their paws, the scraping will produce a visual signal that attracts the attention of other animals to the urine marking or feces.[23]

It is important for animals to be able to distinguish kin from nonkin to prevent inbreeding. Many animals use odorants to accomplish this. Can dogs recognize one another by odors? Puppies that are from four to five and a half weeks old recognize the odor of their siblings and mother, and the mother recognizes the odor of her litter. Mothers are even able to recognize their adult offspring from their odor two years after birth even if they haven't lived with them since they were eight to twelve weeks old. Siblings were only able to recognize one another if they had lived with another sibling.[24]

Odor Markings from Urine

When a dog urinates on a lamppost, it is most likely not just information about gender and species it leaves behind, but also a signal communicating its identity. The urine compounds are volatile and will evaporate after a while, so the marking is diluted and for this reason dogs must leave markings frequently. In mammals, it is usually the males that leave the most markings, and this is true for dogs.[25] Male dogs will lift a leg when marking, something they may also continue doing even if the bladder is empty.[26] A dog that weighs 20 kilos produces 0.5–0.8 liters of urine daily (25–40 ml/per kg).[27] The marking activity is controlled by the hormonal compound testosterone. The production of testosterone begins in puberty and gradually increases with age.[28] This marking activity is therefore not as common in females.[29] The urine of neutered dogs has another odor, and dogs are less interested in urine markings from a neutered dog.[30]

Dogs are especially interested in things like lampposts, tree trunks, and fire hydrants because other dogs have marked them with their own urine. After having sniffed the marking, a dog will very likely leave a marking of its own odor on top of it. Both male and female dogs do this. Dogs compete to be on top. This is perhaps the most well-known and often-used example of what we call scent over-marking, which is also very common in many other mammals. Finding the markings of strange animals will trigger this behavior, especially if the animal is in its home territory. At the same time, marking activity differs greatly from one dog to the next. That is also the case for my dogs. Shib is without doubt a dominant dog. She marks almost everywhere and most of the time. Chilli is not nearly as active. The male dog Tapas, however, marks the most. When we are on a ski trip with Tapas, it is not always an easy matter getting him to pull me behind him on skis the way I want. He is more interested in all the markings along the trail and will stop to mark over these. If Shib leaves a marking, Tapas runs over and marks on top of it. Is he trying to guard his beloved Shib by doing this? Males show less interest in the urine of a female in heat if it also contains urine from another male. This over-marking possibly reduces the chance of another male understanding that a female is in heat, or sends a clearer message that this female already has a mate or is "taken." Usually the markings will only partially overlap. They can therefore also inform females and male rivals about the presence of a male with a high ranking in the proximity of adult females.[31] This over-marking activity is an important means of establishing new relationships without fighting, increasing group belonging, and building up and maintaining stable ranking in the group.[32]

It is said that male dogs can smell female dogs in heat more than four kilometers away.[33] A female dog in heat emits powerful odorants.

In the classic work *The Horrible and Terrifying Deeds and Words of the Very Renowned Pantagruel King of the Dipsodes, Son of the Great Giant Gargantua* from 1532, Renaissance author François Rabelais tells the story of how Panurge took revenge after having been rejected by his beloved. Panurge killed a female dog in heat and cut out her womb. He chopped this up into small pieces, which he then surreptitiously rubbed on the dress of his beloved. Large and small dogs, fat and thin, all arrived bursting with the desire to mate, and sniffed and urinated all over her. Everywhere she went, the dogs followed the trail left by the train of her dress to her home.[34]

It is interesting to observe dogs' marking activity in the wintertime when the markings are much more visible and can be detected as yellow patches in the snow. Do dogs mark over all the other markings, only partially, or beside them? We have also noticed while doing our research on beavers that animals' markings can be yellow or reddish-brown patches that are visible in the snow. When we wanted to find out whether the beaver marked its odor with compounds from castoreum or the anal glands, we collected odor markings made in the snow. Through chemical analyses, it was easy to find out whether they had used one or the other, or both. In the context of our continuing beaver research, we also collect odorants from animals we capture alive. We create two artificial odor markings side by side that the beavers can compare. We record which marking they investigate first, how long they sniff at the different markings, how much time they spend destroying them, and whether they will would leave their own odor there. We can do these experiments with many different combinations of samples.[35] We can put out such odor markings in different beaver areas. The odor can be collected from the beavers' territory or from other areas.

You can also carry out these types of simple odor experiments if you want to determine how your dog reacts to different odor markings. When your dog marks in the snow, you can see clearly where it has done so, since the snow will be yellow. You can collect this odor marking in a disposable rubber glove or a clean plastic bottle. After having collected the marking from your own dog, you can also collect markings from other dogs. If you also have markings from females in heat and females that are not in heat, you can test how your dog reacts to the different markings. If you want to collect many marking samples before you carry out the experiments, you should store these in a freezer and thaw them out before you start. Take these markings to another location and lay them out with a distance of around 15 centimeters between each of them. It is most likely that your male dog will be more interested in the urine from other dogs than in his own. He will also be more interested in the urine from females than males, and more interested in females in heat than those that are not in heat. After having sniffed at the urine, he will probably mark over it with his own urine. In the snow, you can easily see whether he marks beside or right on top of the other marking.[36]

The Puppy Leaves Home

It is stressful for puppies to come to a new and unfamiliar home.[37] The puppy has felt safe and content with its mother and siblings, but at one point it must leave this secure environment. When our puppies left us, we cut up the blanket they had been lying on, so each of the puppies could take a small piece with them to their new homes, and thereby also the odor of their siblings and mother. Puppies usually arrive in a new home when they are between six and nine weeks old. And it is, unfortunately, at this age that they protest the most loudly about having been isolated.[38]Another challenge for the new owner is housebreaking the puppy. A synthetic odorant that is supposed to calm dogs has been developed for this purpose. This substance can be emitted from a container that is plugged into an electric outlet, from a spray, or from a dog collar. The odorant resembles the scent that mother dogs produce when they are nursing their puppies.[39] This substance can have a tranquilizing effect on adult dogs and puppies in stressful situations such as during fireworks displays, when they are home alone, when they are out traveling, while they are being trained, and when they are boarding at kennels.[40] In a placebo-controlled study by animal behavior researchers Katy Taylor and Daniel S. Mills from the University of Lincoln in England, barking in particular was reduced considerably using a dog-appeasing pheromone, and there were also fewer urinations and bowel movements indoors.[41] When dogs are at home alone, the scent of lavender and chamomile can have a relaxing effect on them, while the scent of rosemary and peppermint might make them more agitated.[42] It is, however, important to be aware that different dogs have different preferences when it comes to odors.[43]

Odor from the Anal Sacs

Dogs have anal sacs that are located by the anus. If you think of the anus as a clock face, one of the anal sacs is located between four and five o'clock and the other between seven and eight o'clock.[44] The anal sacs are pouches beneath the skin that are a reservoir for secretions from modified sweat glands (apocrine glands).[45] Normally speaking, the anal sacs are the size of a peanut. The color of the secretion varies from dog to dog and can have a yellowish, grayish, or brownish color.[46] The consistency also varies. It can be watery (it is 88 percent water[47]) or oily, and its odor can be somewhat acrid. The color and consistency from the right and left anal sacs can also vary.[48]

In comparison, the anal sacs of the wolf are surrounded by a layer of muscle that enables them to control when they will deposit a secretion. The secretion is, in other words, not passively deposited with the feces.[49] Cheryl S. Asa, a research scientist in reproductive physiology at the University of Minnesota, led a study in 1985 in which the scientists sedated wolves and colored the secretion in the anal sacs with a dye. The secretion was found in very few of the defecations. In addition, they discovered that the solidity of the feces had no bearing on whether the secretion was deposited or not. The adult male wolves, especially those who were sexually active, deposited secretion more often than females and young wolves. Corresponding studies have not been carried out on dogs, but it is very likely that this functions the same way in dogs as in wolves.

Back in 1969, veterinarian Conrad A. Donovan from Latrobe, Pennsylvania, was curious about how dogs reacted to the secretion from the anal sacs of dogs in heat. He collected secretions from female dogs and presented this to different dogs of both sexes. Adult dogs showed the most interest. When he spread it on the behinds of other dogs, the adult males tried to mate, while the younger males simply followed after them and sniffed. The secretion from females who were not in heat led to sniffing, but not significant or with lasting interest. The same reaction was observed for the secretion from dogs with an anal sac illness. Donovan carried out further trials using the secretion from dogs who had been frightened. When this secretion was applied to the behinds of other dogs, the younger puppies showed an interest, while the older dogs avoided them. It is well known that dogs will spray anal secretion in response to extreme stress, such as when they are visiting a veterinarian. This is an alarm signal that we are familiar with from skunks, who will spray a secretion in self-defense.[50] Chemical analysis of canine anal sacs indicates that the secretion contains information about the individual animal, including its sex and breed.[51]

Many dogs develop problems with the anal sacs. Breeds such as the Chihuahua, dachshund, miniature and toy poodles, and Cavalier King Charles spaniel can have greater problems with their anal sacs than other breeds. Obese dogs can have more problems than thin dogs, but thin dogs can also develop problems. Small dog breeds can also have problems emptying the anal sacs. It is in these cases that the dog starts dragging its rear end in order to try and relieve the pressure. Many also start chasing their own tail or biting or licking themselves in the behind[52] and can have difficulties defecating. If the secretion remains in the anal sacs for a long period of time, it becomes thicker and acquires the consistency of peanut butter, and the sacs become much larger. The color then becomes more brown or yellowish green. When things have progressed to this state, it becomes difficult to empty the sacs and bacterial infections can arise. If a boil should form, this is very painful for the dog. The anal sacs swell up and become reddish, and they can finally burst. The dog must then be given antibiotics and possibly be operated on. Dogs that have such problems should be drained on a regular basis.[53] In the worst-case scenario, a veterinarian will have to surgically remove the anal sacs. It does not always help to drain the anal sacs, because some dogs will recommence behind-dragging activity a mere three weeks later.[54] You can drain the anal secretion yourself, but it is best to find a veterinarian for this task. If you want to try to do so yourself, put on rubber gloves and insert your index finger inside and up the anus (1–2 cm). Using your thumb, take hold of the skin over the anal sac on the outside. Then push upward and inward and toward the opening of the anus. If nothing comes out, the condition may have already progressed too far, or else it could be that you haven't mastered the technique. In either case, you should immediately take the dog to a veterinarian.

The Odor from the Supracaudal Gland

Dogs also have a supracaudal gland (or violet gland) that is located approximately one-third of the tail length from the root of the tail.[55] The gland covers an area that is 2.5–5 centimeters long.[56] There is little known about this gland, but it is said that it is used for odor markings.[57] When I took Shib to the veterinarian Elin Torø at the Telemark Animal Clinic in Gvarv in the autumn of 2013, she was excreting a clear secretion from her tail gland.

Odor Marking from Feces

Dogs can use feces for marking their home territory and particularly where intruders are expected. They deposit the feces in striking and often elevated locations. Markings on elevated sites are easier to detect, in that the evaporation surface area at nose level expands and the elevation protects the marking from being quickly washed away by rain and dew. It has been observed that some male dogs use a "handstand" (standing on their front paws and lifting their behind up in the air) to deposit their excrement in a spot that is higher up.[58] It is most likely that they do this to inform others of their size and thereby also of their competitiveness.

Information from Saliva

When dogs meet, they sniff and lick at the corners of each other's mouths. Are they acquiring information from the odorants of the glands or from the saliva, or possibly from both? It is interesting to note that male dogs sniff and lick the corners of the mouths of females more frequently than of males.[59] Dogs also lick each other's fur and thereby transfer saliva secretion. A mother will do this with her puppies to clean them and to stimulate urination and defecation.[60] The puppies also transfer their saliva to the mother's nipples and possibly also leave markings on the nipples in this way. Saliva contains high concentrations of hormones such as progesterone, testosterone, and cortisol, making it possible to sniff out the differences between the sexes.[61]

Information from Sniffing at Ears

Dogs sometimes sniff at each other's ears when they are greeting one another. Earwax—a yellowish gland secretion—is produced by modified sweat glands. The ears are an excellent area for bacteria and other microorganisms to thrive, and these can produce different odors. In 1977 veterinarian and behavioral biologist Ian F. Dunbar at the University of California, Berkeley, found that male dogs showed little interest in earwax if it was from other males. If it came from females, they were a full twenty times more interested.[62] By sniffing at each other's ears, at the very least they can acquire information about the other dog's sex.

5

Search and Rescue

On March 6, 2011, there was an avalanche in Lifjell, Bø municipality in Telemark, Norway, triggered by an arctic dog named Fenris. His owner, Per Tore Iversen—who also had a female dog, Isa, with him—plummeted down the mountainside. His descent came to a halt at the end of the 300-meter-long avalanche, where the snow was seven to eight meters high. Iversen was fortunate. He could breathe but was buried in such a way that only his head was sticking up out of the snow. His skis were lying in a cross over his stomach. Even though there were steel rims on the skis, he had not been injured. Luckily, Iversen could also move his left arm. He managed to remove his glove so he could clear away some of the snow. In this way he freed his hand and was able to reach the shovel in his knapsack. The snow was as hard as a rock, and he had to "chop" his way out with the shovel. Isa had managed to run out of the way of the masses of cascading snow and was therefore not caught. After the avalanche stopped, she lay down directly behind Iversen. Fenris, on the other hand, was nowhere to be seen. Isa was not interested in searching through the avalanche. Iversen called and searched as best he could, but he did not find Fenris and finally had to give up. The Rescue Dog organization and a Red Cross group also took part in the search for Fenris, but they were not able to find the dog.

Iversen was unable to accept that his best friend had not been found, and the next day he took Isa back to the area. This time he was accompanied by his buddy Gunnar Fagerli and his two Alaskan malamutes, the female Atka and the male Nanuq. Nanuq and Fenris knew each other well and would often engage in playful tussles. Immediately upon reaching the avalanche site, Nanuq began digging furiously in the snow. Gunnar and Iversen understood immediately what had happened and they, too, started to dig. Nanuq had started digging just two meters from the place where Iversen had been stuck the day before. And there—under a meter of snow—a damp snout came into view. Fenris was lying there all curled up and he was alive. The only injury he had suffered was inflammation in his tail, but after treatment with cortisone, he was as good as new. Fenris had not suffered any other injuries after spending 28 hours under the snow. After Fenris was dug out of the avalanche, he jumped right into Iversen's lap and peed for a long time. He had not urinated while buried in the snow. For dogs it is very important to have a clean living space (den), and that is quite possibly why Fenris had not urinated.[1]

Fenris hitched a ride a short distance from the mountain with the Red Cross snow scooter patrol, but after a little while he jumped off. He preferred to walk down together with his two-legged and four-legged friends. Fenris has since returned to the avalanche site and does not appear to have been affected by his time spent buried in the snow. But why wasn't he found by the rescue crews and their dogs? The search and rescue dogs had been trained with human odors, and very likely it was for that reason they did not find the dog. Fenris's playmate Nanuq, on the other hand, was of course interested in the scent of his playmate. He was ready for another lighthearted tussle. Another possibility is that the snow conditions changed the following day, allowing more odors to rise to the surface.[2]

Search and Rescue

The command "search and rescue" (SAR) is also used to describe dogs employed for these purposes. The cues the dogs are given vary considerably and other phrases, such as "track" and "go find," are also used. Dogs were used on rescue missions in the eighteenth century and possibly even a hundred years before that.[3] Search and rescue dogs are often classified according to the environment in which they work; dogs that track using air scents are used to searching for missing persons in the wilderness, in avalanches and landslides, in both collapsed and intact buildings, in rivers, lakes, and many other places where people have been reported missing.[4] They are also used to search for cadavers (see chapter 7, "Police Work"). Searches can be broken down into seven main types: (1) woodland searches and searches in large land areas, (2) searches for evidence, or in small areas, (3) searches for human remains/cadavers, (4) water searches for drowning victims (see chapter 7, "Police Work"), (5) avalanche searches, (6) searches for survivors following a disaster, and (7) searches for human cadavers after disasters.[5]

Most search and rescue dogs are medium-size breeds, such as golden and Labrador retrievers, Belgian sheepdogs (the Malinois, the Tervuren), German shepherds, or border collies. The larger breeds are usually used for avalanche searches, while smaller breeds are used to search in rubble. It is important that the dogs are motivated, calm, and social when in the company of other dogs and people, and that they don't have any physical infirmities. It is also important that they are not aggressive. These dogs are usually owned, instructed, and trained by volunteers who are not paid to do this job.[6] Search and rescue dogs work both on and off a leash. They can work with their nose to the ground, where trailing dogs are usually presented with the scent of an object from the person they are supposed to find before the search begins (scent discrimination trailing).[7] Alternatively, search and rescue dogs can be presented with the scent of a place where one knows the person in question has been for a period of time. A scent article can also be stationary, such as a car seat, window, or door handle. Hairbrushes, toothbrushes, eyeglasses, and especially used tampons are good scent objects. And naturally, the better the scent article, the better the chances of the dog finding the person.[8] A search and rescue dog taking its cue from air scents does not work on a leash and can move across large areas of land. These dogs are more flexible and are different from the scent-discriminating trailing dogs in that they need not sniff an object before starting the search.

How long an odor lasts varies, and many different factors have an impact on this. It can be a matter of different environmental factors, such as the temperature, rain, wind force, and direction (see chapter 6, "On the Hunt"). It is easiest to find the person being searched for after an hour or two, and it becomes much more difficult as time passes. According to policeman and author Jeff Schettler, after twelve hours there is a less than 15 percent chance of finding the missing person. If the trail is more than twenty-four hours old, the probability of finding the person is less than 10 percent. Few dogs are able to follow a trail that is more than twelve hours old, and it is best to follow the trail within one to four hours.[9] However, there are reports of dogs that have succeeded in following a 105-hour-old trail—in other words, a trail almost four and a half days old.[10] The longest trail we know of that has been tracked successfully was 216 kilometers long.[11] Of course, the likelihood of finding the person being searched for also depends on where the person is found (the terrain), the type of person and their activity level (see chapter 3, "A Good Judge of Character"). The training the dogs have received is also significant: it is important to train dogs on both new and old trails.

The Bloodhound: "The Ferrari of Noses"

Bloodhounds are often associated with searches for people. Unfortunately, only a limited amount of research has been carried out on the ability of the traditional bloodhound to track and distinguish between the odors of different people,[12] although in folkloric accounts it has been considered the best dog; it is often said to have "the Ferrari of noses."[13] The bloodhound was bred to track people using its sense of smell by French monks who bred the St. Hubert bloodhounds in the seventh century.[14] In the 1830–40s, bloodhounds were used by the military to track down Indians and runaway slaves in the southern states of Florida and Louisiana.[15]

Although the bloodhound's exceptional sense of smell has been described in oral accounts throughout history, this breed's ability to track a human being was not scientifically tested until 2003.[16] Eight bloodhounds—three inexperienced dogs that had received less than eighteen months of training before the test and five experienced dogs that had received training for at least eighteen months—were put on five different trails of lengths ranging from 0.8 to 2.4 kilometers. The trails had been created forty-eight hours earlier and led to a fork, where the dogs had to decide whether to follow the trail to the right or left. One of the trails was from a person for whom they had received an odor sample before beginning the search. The trails went through a local park, a university campus, and finally a city. All these places were heavily frequented by human beings. The results showed that 78 percent of the eight equipages completed the tracking test and identified the right person. The inexperienced dogs did not do as well (53 percent) at tracking as the experienced dogs that had also been used in searches previously (96 percent). Training and thereby experience was accordingly very important for a successful search.

In another study done in the United States, the FBI tested a bloodhound's ability to discover human odor in a populated residential area. They selected a person who had lived in a house in Stafford, Virginia, for seven years before moving to Albuquerque, New Mexico. Six months after the person had moved to New Mexico, research scientists from the FBI placed a bloodhound at an intersection several houses away from where the person had lived in Stafford. The dog subsequently was allowed to sniff a letter the person had sent from New Mexico, and it then went directly to the house where the person had lived. In this case, the traces of the person's odor in the house had been preserved in spite of being exposed to the elements for more than six months.[17]

It is a myth that bloodhounds will easily lose the trail of a person if he or she goes into the water. Most of us have seen old black-and-white movies in which the only route of escape for a fugitive is a river.[18] Many criminals copy this strategy, but running in a river does not help. Some bloodhounds in fact work better around water than in other locations. Body secretions are almost like oil on the water surface and flow with the wind and current. The odor can also attach itself to vegetation, stones, and soil along the water's edge and can spread several kilometers downstream.[19]

Disaster Search Dogs

The disaster search dog searches for people who are trapped in rubble after a building has collapsed.[20] This type of search dog also takes part in searching for people in areas struck by other disasters, such as landslides and rockslides, floods, earthquakes, tornados, hurricanes, tsunamis, mining and railroad accidents, and military attacks.[21] While other search and rescue dogs are used for searches in large areas—such as forests, lakes, or mountain regions—disaster search dogs are used in smaller areas. Searches of rubble were first implemented in England during World War II, when the Civil Defence forces used dogs to search through the rubble following German bombings.

Disaster search dogs were used after the tragic bombing of the Alfred P. Murrah Federal Building in Oklahoma on a spring day in 1995, and after the terrorist attack on the World Trade Center and the Pentagon on September 11, 2001.[22] Following the latter terrorist attack, 250 to 300 rescue dogs were used.[23] The remains of the buildings were moved to a large area, where the dogs searched through tons of rubble for human remains, such as tissue, bone, or even body fluids. One dog, for example, gave an indication for a small object that looked like a wood chip, but which turned out to be a piece of a human rib bone.[24] Dogs were also used in Lockerbie in southwest Scotland in 1988, where an airplane crashed due to a bomb on board. The explosion killed 243 passengers, 16 crew members, and 11 people on the ground.[25] After the earthquake in Friuli, Italy, in 1976, 12 dogs were able to find 42 survivors and 510 fatalities.[26] After a tsunami struck Yamada, Japan, in March 2011, dogs were flown in from all over the world.

The disaster search dog can search for both living and dead people. There is little consensus regarding whether dogs should be trained for both, because of the risk of their giving an indication for a dead person instead of a live one. A study done in 2006 by Professors Lisa Lit and Cynthia A. Crawford at California State University, San Bernardino, concluded that dogs that were trained for both living and dead human beings should not be used when the smell of corpses was present and the search is for survivors.[27] The survivors might be overlooked, which can lead to any injuries being compounded or the loss of lives. Disaster search dogs should also be trained so they don't give indications for food, live or dead animals, or clothing.[28] A healthy, adult person who is uninjured and has access to fresh air has a large chance of survival if found within 72 hours. Approximately 80 percent of those who are found alive are rescued within 48 hours. After 72 hours, the survival rate quickly drops and without access to water few people can live for longer than 120 hours, in other words, five days.[29]

A trained rubble search dog will be able to quickly search through the upper layers of a collapsed building and has a large possibility of finding survivors. The dog sniffs its way to the odor of human beings' breath, blood, skin, and urine. Carbon dioxide, ammonia, and acetone are reliable indicators of an active metabolism, and these compounds will spread quickly through collapsed buildings.[30]

In general, the disaster search dog is not given an object with a human odor to sniff, because such an article is normally not available. These dogs sniff the air surrounding the area where the search for people is to be done (air scents). The dog does not work on a leash, because it can search more effectively alone in collapsed buildings. The work at a disaster site can be extremely dangerous both for the dog and the dog handler. When the dog finds the scent of humans caught in the rubble, it moves toward the area with the strongest scent. It can detect the scent of humans found many meters inside a building. When it can't get any closer, it will begin to bark to indicate that it has found a living human being in the rubble. This location will then be marked, and the dog will receive a reward, such as a game of fetch with a ball. If there is a suspicion that there may be more people found in the rubble, the same dog is used again, or several equipages will work in shifts. If there are no survivors, the handlers will often have volunteers hide in the building so the dog can find them. For the dogs finding survivors is a game, and the dog's interest in the game must be encouraged, upheld, and reinforced. The handler must be positive and enthusiastic with the dog, even if the outcome of the search is negative.[31]

Disaster search dogs are being outfitted with more and more advanced equipment.[32] This can entail everything from dog eyeglasses with cameras to different types of harnesses with tummy bags, containing medical supplies, a radio, food, and water.[33] Since 2008, dogs in the United States have been equipped with accelerometers (also found in smartphones), which transmit wireless signals to a monitor where the handler can study, among other things, the dog's body position.[34] For example, the dog may have been trained to sit when it has found a dead human being and to stand and bark if the person is alive. The dog handler will thereby know what the dog has found even if it cannot be seen or heard. Research scientists can gauge the dog's physiology by using a special sensor.[35] This sensor is capable of measuring the oxygen saturation in the brain by illuminating the brain tissue in the skull and recording its reflection.[36] The research scientists can also measure how much energy the dogs expend on different activities with the help of accelerometers.[37] In 2010 yet another useful tool was invented for the disaster search dog. Professor of computer science Alexander Ferworn at Ryerson University in Toronto and his research team developed a robot snake that the dog carries in a belly pack. The robot snake is freed when the dog finds the person it is searching for in the rubble and the dog sounds the alert by barking. The rescue team can thereby see and speak with the person trapped in the rubble.[38]

Can Dogs Give Early Warnings for Earthquakes?

Many wild and domesticated birds and mammals can be sensitive to signals preceding an earthquake.[39] It has been reported that dogs react up to twenty-four hours before an earthquake. Can these signals be odors? And can dogs be trained to notify us before the earthquakes occur? If so, volatile sulfur compounds are the most probable odor candidates, since sulfur oxide is emitted by earthquakes.[40]

The Mountain Rescue Dog

The monks at the St. Bernard monastery in the Swiss Alps used St. Bernard dogs to search for missing persons in the mountain passes between Switzerland and Italy. The St. Bernard pass is located 2,438 meters above sea level. In the course of two hundred years, more than two thousand people were rescued in the St. Bernard pass, from young children to Napoleon's soldiers.[41] The first avalanche search and rescue dogs were put to work in this pass searching for avalanche victims as far back as the seventeenth century.[42] From 1800 to 1812, the well-known St. Bernard dog Barry operated in the region. He saved the lives of forty-one people who had been caught in an avalanche or had gotten lost in the mountains between Switzerland and Italy during the winter.[43] Between 1816 and 1818, there were heavy snowstorms that took the lives of many dogs in avalanches while searching for people. The last rescue done by a St. Bernard in this area was in 1897 when a twelve-year-old boy was found alive but chilled to the bone. The St. Bernard dog was formerly a muscular, wiry-haired dog. Dogs from the Newfoundland and English mastiff breeds have been bred into the St. Bernard of today. This has made the St. Bernard dog both large and very heavy.

We often read in the newspaper or hear on the news stories of how search and rescue dogs have found people buried under the snow. Today dog breeds such as the Australian shepherd, border collie, German shepherd, rottweiler, giant schnauzer, Nova Scotia duck tolling retriever, golden retriever, and Labrador retriever are used for avalanche searches.[44] A mountain rescue dog can find people buried several meters under the snow.[45] In Austria a dog found a person 7.3 meters deep in the snow.[46] And dogs are efficient. A rescue crew of twenty people will need twenty hours to search 0.625 acres. A dog, on the other hand, can search through the same area in the course of a mere two hours.[47]

It is naturally very important to find people who are buried in the snow as quickly as possible. The chance is greater than 90 percent that the person being searched for will still be alive within fifteen minutes, but this drops to 30 percent after thirty-five minutes. After this, it is impossible to survive without an air pocket.[48] If there is an air pocket in which to breathe, an avalanche victim can survive for up to twenty hours.[49] Finding the person can be a challenge. It is easier to find the person if the avalanche has moved across short distances (and elevations) or if the snow is very loose.[50] An avalanche can travel at speeds of up to 193 km/hour (56.5 m/second) and carry several tons of snow.[51] The characteristics of the snow and its density determine whether odor slips through the snow and up to the surface and how the odor moves through the snow. The odorants from a person who is buried in the snow follow channels in the snow where the density is the lowest. An ice cap around a person in the snow will cause the odorants to follow paths other than straight up to the surface. The deeper in the snow the person is lying or the denser the snow, the longer it will take for the odorants to reach the surface.[52] In powder snow, the dispersal rate of odor can vary from one minute per meter or less, to fifteen minutes per meter if the snow is wet.[53] In some cases, the snow can be so compact that the odor will not slip through until holes have been drilled in the snow. Dogs can easily search in temperatures as low as –30°C. The stronger the wind, the greater the likelihood that the dog will give an indication on the sheltered side of the location where the person is buried. Since the air moves constantly through the snow, the wind will also affect the odor beneath the surface of the snow.[54] It is important to be aware that in some cases odorants can rise to the surface of the snow up to 15–20 meters away from the avalanche victim.[55]

6

On the Hunt

In the course of almost forty years of using pointing dogs for ptarmigan hunting and research, the now-retired professor of ecology Howard Parker has experienced numerous examples of the dog's incredible ability to sniff out game under extremely difficult air-scenting conditions. In mid-November 1988, Parker was out willow ptarmigan hunting with his dog Balder, a Brittany spaniel, in a narrow mountain valley. The sky was clear and the temperature was −3°C; there was no wind and approximately 25 centimeters of new snow on the ground. They were walking through the valley when Balder froze and pointed, and two willow ptarmigans took flight. One of them fell after the first shot, while the other flew ahead into the valley. Parker suspected that the second shot had also hit its mark, so they performed a thorough search without any results. Parker continued hunting, and at sunset both he and his dog were quite tired. Parker unloaded his gun, called in Balder, and began walking back along a logging road running through the bottom of the valley.

After having walked for about a half hour, and just below the location where they had earlier that day searched for the wounded ptarmigan, Balder suddenly stopped, pointing his snout up the mountainside. There was no wind whatsoever, but the dog had set with such conviction that Parker reloaded his gun. He gave the order to advance, and Balder began moving warily up the hill. After repeated orders to advance across some 30 meters without finding the bird, Parker started losing faith in Balder's indication. However, the dog remained certain, so they continued climbing up the slope. After about 10 more meters, Balder refused to continue and dropped his nose to the ground. The light was poor, but Parker could just make out a slight hollow in the surface of the snow directly in front of the dog. A snow roosting ptarmigan, he thought, and lifting his shotgun, he stomped hard on the ground to flush it out. But nothing happened, so he dug carefully into the hollow with his right hand. To his astonishment, a dead and half-frozen ptarmigan lay there, around 10 centimeters beneath the surface. A small spot of blood on its chest indicated that this was the same ptarmigan he had been searching for earlier that day. It had clearly been hit by a single shot and had managed to fly away.

Balder had detected the scent of a dead bird from a distance of approximately 35 meters, a bird that had been buried beneath around 10 centimeters of snow for five to six hours. How could this happen? When the sun is shining, the snow surface and surrounding vegetation absorb the energy from the sun. The temperature of the snow surface will then rise slightly, and the air stratum just above the ground will also become warmer. The heated-up stratum of air will also have a tendency to rise, because warm air is lighter than cold air. When the sun goes down, the energy supply to the ground will stop. The ground will nonetheless continue to radiate heat (due to long-wave radiation), and both the snow surface and the stratum of air just above it will cool down. If there is little wind and a sloping terrain, a thin, cooled (and therefore heavy) stratum of air will flow slowly down the hill. It was this thin air stratum that had transported the odorants from the dead ptarmigan buried in the snow and that were detected by Balder's incredibly sensitive nose. The only one who was disappointed about this find must have been the red fox who had been cheated out of a ptarmigan for its supper.[1]

Different Kinds of Hunting

For thousands of years, humans have enlisted the help of dogs when hunting for food. The use of dogs in hunting is also believed to have motivated their domestication (the taming and breeding of animals so as to enable them to adapt to life with human beings) since this made hunting more effective.[2] The hunters encountered different kinds of prey.[3] There is a painting on the wall of a cave from approximately 10,000 BC in Cueva de la Vieje in Spain that allegedly depicts a dog hunting deer.[4] In the Middle Ages, the use of the Irish wolfhound in hunting wolves and Irish giant deer was well known, but this breed is no longer used in Europe. In this form of hunting called coursing, the dogs first locate their prey by sight, then chase it, capture it, and kill it. In Kyrgyzstan, this dog breed is still used for wolf hunting.[5] The saluki, the royal dog of Egypt, was used along with falcons for gazelle hunting.[6] Red fox hunting with dogs (coursing) started in Great Britain in the seventeenth century[7] and is also practiced in Australia, Canada, France, Ireland, Italy, Russia, and the United States. Hunters follow the dogs on foot or on horseback. The sport is controversial, especially in Great Britain, and was prohibited in both England and Wales in 2004.[8]

A hunting dog can work in many different ways. The dog can track or trail game with or without baying (trailing breeds or hunting dogs on leashes), set and point toward game (setters and pointing dogs), find and flush out game from above (flushing dogs) or below (burrow hunting dogs) the ground, or find and retrieve shot game (retrievers).[9] Many of us are familiar with the use of dogs in hunting for large-game animals such as moose, roe deer, red deer, and reindeer.[10]

Hunting with a dog is particularly advantageous when the prey density is low.[11] In 2013 research scientist Christopher Godwin at Trent

University in Peterborough, Canada, and his colleagues found that hunters who had one or more dogs with them increased their hunting success for the white-tailed deer by 26 percent per hunting day. They also shot more adult bucks and fawns compared to hunters who were not accompanied by dogs. If there were also dogs in a nearby neighborhood (less than two kilometers away), the hunt was not equally successful.[12]

Many people use dogs to hunt small game, such as grouse, hare, rodents, and a variety of carnivores such as red fox and martens.[13] The hare hunting dog sniffs its way to the nest-like depression in the ground (form) that is the hare's daytime lair, startles the hare out, and then pursues it (beagling).[14] It is probably less common knowledge that we also use dogs on burrow hunts for rabbits, red foxes, badgers, raccoon dogs, and beavers.[15]

In Latvia, the West Siberian Laika is used to hunt beavers. Beavers live in families normally made up of an adult couple and their young from the two preceding years. They build lodges and dig out bank burrows. Here they live a life relatively protected from predators. The entrance to their home is usually underwater, made possible by the dams they build, so it is difficult for predators to reach them.[16] In Latvia, beaver hunting is done using lethal traps and with firearms. But there is no doubt that the most effective form of beaver hunting is with dogs. When three to four traps are used, the hunters will trap approximately one beaver a day, and they will shoot one beaver about every third day. When the hunters bring dogs along with them, the take will be up to four or five animals a day; normally they will shoot all the animals in a family group. The first thing the hunters do is to tear down the dam (this is prohibited in many places without a permit from the county wildlife authorities), so all the water runs out of the beaver dam, exposing the underwater entrance to the beavers' home. The dogs are then sent in. Their keen sense of smell enables them to determine quickly where the beavers are hiding in the lodge. The dog chases the beavers out of their home to outside, where the hunters are waiting with their rifles. The beaver has no chance of escaping alive.[17]

TABLE 2: Some of the more common and less common animals hunted, the types of hunting dog groups, and the most common breeds for each form of hunting

Animal hunted	Hunting dog groups	Breed
Birds[a,b]	Pointing breeds	English setter, Irish, and Gordon setters; German shorthaired, wirehaired, and longhaired pointers; Münsterländer; and Brittany spaniel
Birds[b]	Flushing breeds	Springer spaniel, cocker spaniel, and retrievers
Birds[b]	Treeing breeds	Finnish spitz
Birds[b,c]	Retrieving breeds	Labrador, golden, flat-coated, curly-coated, Nova Scotia duck tolling, and Chesapeake Bay retrievers
Roe deer[d]	Trailing breeds	dachshund, Drever, beagle, basset breeds (grand basset griffon Vendéen and Norman Artesian basset)
Roe deer[e]	Flushing breeds	German spaniel and Jagdterrier
Moose[f]	Trailing breeds/ hunting breeds on leash	Norwegian elkhound (gray and black), Swedish Jämthund/Swedish elkhound, Swedish white elkhound, Karelian bear dog, West Siberian and East Siberian Laika, Hällefors elkhound, Finnish spitz, and Norrbottenspets
Red deer[g]	Trailing breeds	dachshund, Drever, and basset
Red deer[g]	Hunting breeds on leash	Gray and black elkhound and Swedish Jämthund/Swedish elkhound
Red deer[g]	Flushing breeds	German pointer, terrier, and setter
Hare[a,h]	Trailing breeds	Dunker, Hygenhund, Hamiltonstövare, Halden hound, Schweizer Laufhund, Lucerne hound, Finnish hound, beagle, Drever, dachshund, and basset

Fox[a]	Trailing breeds	Schiller hound, Hamiltonstövare, and Hygenhund
Fox[a,i]	Burrow hunting breeds	dachshund and terrier breeds
Badger[j]	Burrow hunting breeds	dachshund and terrier breeds

[a] K. V. Pedersen, Småvilt- jakt of fangst (Tun Forlag AS, 2008).

[b] J. B. Steen and E. Wilsson, Fuglehundens ABC & D (Hundskolan i Sollefteå AB, 1993).

[c] S. Christoffersson, Bird dogs, dressur og jakt (Schibsted Forlag, 1999).

[d] R. Andersen, A. Mysterud, and E. Lund, Rådyret- det lille storviltet (Naturforlaget, 2004).

[e] K. V. Pedersen, Rådyrjakt med hund (Tun Forlag, 2007).

[f] K. V. Pedersen and J. Unsgård, Elghunder og elgjakt (Tun Forlag, 2010).

[g] E. L. Meisingset, Hjort og hjortejakt i Norge (Naturforlaget, 2003).

[h] J. C. Frøstrup, Hare og harejakt (Teknologisk forlag, 1996).

[i] D. Grandjean and F. Haymann, Hunde-encyklopedi (RK Grafisk, 2010).

[j] S. Christoffersson, Jakt i Norden (Cappelen Damm AS, 2008).

How Do Dogs Find Game?

In general, we can say that dogs hunt game in the same way that they track and search for us. Dead skin cell flakes carrying skin odor, or that have been altered by bacteria, are also constantly shed by birds and mammals as they are being hunted.[18] The scent of game can also stem from the breath, hair, feathers, genitals, rectum, glands, or body fluids.[19] Depending upon particle weight (molecule weight) and the wind, odorants are usually deposited from a few meters to 20 meters or more from the trail.[20] Game will leave behind footprints, or the animal's body odor is left behind on vegetation or other surfaces around the trail. A footprint also causes the release of fluids from plants, and bacteria break down the dead plant cells above or in the ground.[21]

Dogs often move in a zigzag pattern toward the source of an odor. As the dog approaches the source, the odor becomes fresher and more concentrated.[22] The game is able to reduce their emission of some odors, such as sex hormones, thereby making it more difficult for the dogs to find them. They can inhibit the emission of other odorants but not stop them completely. However, there are some odorants, such as those produced by anaerobic bacteria, over which game have little control.[23] In the case of roe deer or fawns, metabolic odorants must be emitted, but these animals do have the ability to inhibit this to a minimum. There are many reports of dogs that have stood more or less on top of roe deer and red deer fawns without noticing them. A red deer can reduce its breathing and heartbeat by 50 percent when frightened by a predator, while the brooding willow ptarmigan can present the dog with even greater challenges, because they can reduce their breathing and heart rate considerably, for up to 20 minutes.[24]

In general, human beings detect odorants at one and a half to two meters above the ground, since for most of us that is our nose level. It is easiest for us to detect light odorants with a low molecule weight (less than 300 daltons) and high volatility. The dog's nose level is lower than that of humans. Dogs find game by tracking with their nose close to the ground, or they pick up on odorants in the air by lifting the nose (air scenting). Some employ a combination of these techniques. Dogs are able to detect odorants with a much higher molecule weight and lower volatility than humans, since their nose is closer to the ground. This ability to also detect the heavier compounds is advantageous for tracking game, because it is more likely that these odorants will be found on a trail than the lighter, more volatile odorants.[25] When dogs do searches by air scenting, it is the lighter, more volatile compounds they detect.

Different Factors That Will Influence a Hunt

Many factors influence the amount of odorants deposited on the ground and in the surrounding environment, and thereby also the dog's hunting success. The dog's olfactory range and ability to track a specific kind of game will be influenced by the following:

- the species being hunted
- the age/size of the game
- how old the trail is
- the game's speed of movement or if it is stationary
- if the game is stressed or injured
- if the game is sick
- the sex of the game
- the diet of the game
- vegetation type and the terrain in which the game is moving
- the ground surface[26]

How long a particular trail will last varies according to the species leaving the trail. This is among other things due to the fact that different kinds of game move differently (birds fly from place to place, while mammals walk or run), or they have different types of glands.[27] The odor from the interdigital glands in the roe deer's hooves makes it easier to follow its trail than, for example, the trail left by a hare.[28] The glands are not always full grown before the animal has reached adulthood, and adult animals often have a stronger odor than younger animals.[29] If the game is lying still, it emits much less odorants than when it is in movement. If the game is in movement, the body temperature increases and more odorants are produced. Deer fawns that are sick and have diarrhea, a high body temperature, or are secreting mucus or blood produce a stronger odor than healthy fawns.[30] The sexes have different odors, and males often have more and larger scent glands than females; also the bodies of males are usually larger than the bodies of females and therefore give off more odorants.[31] Small game can also be difficult to track because a small body size implies less odor emission.[32] The odor emitted is also influenced by what the game has eaten.[33]

The type of terrain in which game is moving about is decisive to the success of the hunt. A newly plowed field is a difficult place to track since odorants from the soil can cover up the scent of game. Asphalt, sandy beaches, and gravel roads retain few odorants from game. A slippery or hard surface contains fewer binding sites for odorants, so these disappear more quickly. On stones and concrete, there is not much bacteria growth. Warm sand will usually kill the bacteria on skin cell flakes and they will dehydrate. Dry sand and gravel have little or no bacteria activity and little or no vegetation. If a red fox should get clay on its paws, the amount of odorants deposited on the trail will be reduced because the paws will not come into contact with the ground. Dogs can thereby lose the trail of a fox when it is moving across a recently plowed field. More lush vegetation will contain more bacteria and odorants. The vegetation provides shade and moisture and bacterial activity is thereby increased. In order for the bacterial activity to continue, there must be a little bit of moisture. A very warm and dry atmosphere will dry out the bacteria's nutrients (skin cell flakes and plant cells). Early in the morning, dew forms on vegetation, which rehydrates the cells and replenishes the food source of the few remaining bacteria. It is much easier for a dog to detect the scent of game that has passed through bushes or tall grass than game that has walked across a grazing area. Coniferous forests are difficult tracking grounds since there is little underbrush. The pine needles on the ground also have an aromatic scent that can disguise the scent of game. Some plants can also emit a scent so strong that in some cases it covers up the scent of game. Hunters themselves can also increase the scent released by the vegetation as they walk through the terrain. If two dogs are hunting together, they may start to compete, causing them to run past the source of the odor.[34]

Does It Help to Wear Scent-Proof Clothing While Hunting?

Scent-proof clothing is designed to prevent game from detecting our odor, but does it actually work as the manufacturers claim? In 2002 wildlife management specialist John A. Shivik at the National Wildlife Research Center in Fort Collins, Colorado, investigated whether his seven dogs could find people wearing scent-proof clothing. In the course of 42 trials, only 1 out of 21 people wearing scent-proof clothing was not found, while all 21 people wearing ordinary clothing were found. The time the dogs spent on finding these people was very similar for both groups (an average of 3.4 and 2.7 minutes, respectively).[35]

The Impact of Weather Conditions on Hunting

Game can use the weather conditions to make it more difficult for a dog to follow their trail.[36] The success of the hunt is affected by the wind, the temperature and the time of day (sunlight), and whether there is humidity and rain or turbulence and snow. In general, high temperatures cause more evaporation and bacteria activity and thus more odors. But temperatures that are too high will kill bacteria and stop the production of odors.[37] As air becomes warmer, it also becomes lighter and the odorants can thereby ascend instead of dispersing at the dog's nose level.

On land, odor generally behaves in the same way as smoke. Over time, it spreads in a conical shape from the source. The wind direction determines the primary direction of the cone's axis, but whether the cone axis remains close to the ground or rises is determined by the temperature distribution from the ground and up into the atmosphere. If it is warmest along the ground, the cone's axis will rise and the odor will move upward. On the other hand, if it is coldest along the ground, the odor will stay at the lowest level. How quickly the odor spreads out and away from the cone axis ("is thinned out") depends on the wind speed and turbulence (eddying), which in turn depends upon the temperature, wind speed, and the unevenness of the terrain. A lot of wind means a rapid thinning out of the odorants. On the other hand, if the wind conditions are calm, local temperature variations will play an important part in the dispersal of odorants. Under such conditions the variations in the temperature will also contribute to creating local winds, which follow the terrain and can be used by hunters.

The Spreading of Odor Due to Wind

Everyone who hunts with a dog knows that the wind is critical for a successful hunt. Hunters know that the wind must be blowing toward them and away from the game if they are going to come within shooting range. The dog then has an easier time detecting the scent of the game, since the odorants are blown toward it, while the game is unable to smell the hunter and the dog. The body's airstream can carry some odorants against the wind, but the odorants do not move far when there is a headwind and will not reach the game if it is a few meters away. Under favorable wind conditions, the dog can detect a deer several hundred meters away.[38] If the wind speed is low (and therefore with little turbulence), the odorants will not spread as far from the trail (or cone axis). Bird dogs have difficulties finding birds when the wind speed is more than 10 km/hour (2.8 m/second). The ideal wind speed is between 3 km/hour (0.8 m/second) (light air) and 10 km/hour (gentle breeze).[39] Many hunters will wet one finger and hold it up in the air in front of their face to determine the wind direction. However, it is important to remember that the conditions farther down at ground level can be completely different. A better method is to observe the campfire smoke against the terrain[40] or to use a lightweight scarf/ribbon.[41]

The wind at the top of a cliff can create "dead zones." Game will seek out such dead zones for protection.[42] It is easiest to see how the wind moves around cliffs and summits in the wintertime, when light snow will be blown around and deposited on the terrain, such as in the formation of snowdrifts.[43]

The atmosphere's stability depends on temperature variations at different altitudes and therefore usually changes in the course of the day. The atmosphere is more stable and the wind calmer at night than in the daytime. In terms of odors, it would be better for a dog to hunt at night because the concentration of odors close to the ground is higher then.[44]

The Temperature and the Time of Day

In general, we can say that the best time to hunt is at sunrise, at night, or at sunset when the air is calm and the ground is damp (from dew or light rain). The worst time for a dog to track game is in the middle of a sunny day or when it is raining heavily.[45] If there is a lot of sunshine with high temperatures, odorants are broken down more quickly or they evaporate rapidly so the trail does not last. Such conditions will also kill the bacteria on the game's skin cell flakes or feathers.[46] Cold air is heavier than warm air, and the air temperature changes far more quickly and more dramatically than the ground temperature. When the sun goes down in the evening, the rays from the sun stop shining on the ground and warmth emanates out of the ground. This causes the ground surface and air stratum just above the ground to cool down quickly. The cold air, containing odorants from the game or its trail, seeps downward where the ground slopes to lower-lying areas (like water). It is therefore best for the dog to approach from below on the terrain in the evening and until sunrise.[47] When the ground is heated up by the sun in the morning, the air stratum above the ground becomes warmer and the air will flow upward to higher-lying areas.[48]

Then it is best for the dog to approach from above on the terrain, as this makes it easier for the dog to detect the air containing odorants from the game or its trail.[49] When, for example, a bird's odor cone is located above a low-lying cold air mass, the odor will not descend to the ground. A bird sitting in a tree or on a mountaintop will therefore not be detected by a dog.[50]

Local variations in the landscape often create differences in temperature, which in turn create local winds such as "mountain breezes and valley breezes" or "sea and land breezes."[51] If the contact of sunlight with the ground in a given area is irregular, the surface receiving the most sunlight, such as on southern slopes, will become warmer in the course of the day than other surfaces. The air above these surfaces will be warmer than the air above colder surfaces, such as northern slopes. The air above southern slopes will therefore rise and be replaced by air from northern slopes, which is colder. Birds found on southern slopes are thus more difficult to detect than those found on northern slopes. It is warmer on southern slopes, so there is a lot of upslope flow.[52] On sunny days the temperature in the woods is lower than outside, and one of the reasons for this is because the treetops prevent the sunlight from reaching the ground and warming it up. The temperature variations can produce local winds that move along the ground in the forest and to adjacent areas and clearings. At night the opposite is the case, with the ground in the forest warmer than in the adjacent clearings. In the course of the day and early in the evening, it is therefore advantageous for the game being hunted to be on a southern slope or on a dry surface so its odor is transported up into the atmosphere and thereby above the dog's sniffing zone. In the evening, the game will usually move to places with nocturnal upslope flows, such as wet or densely overgrown areas. These areas will remain warm longer than other areas and produce nocturnal upslope flows.[53] Local winds not only arise where the ground slopes, but also rise between land and water surfaces with different characteristics. Water is warmer than land at night, so airstreams arise on land and move out across the water surface during the night, when the stronger winds have calmed. Game will therefore often be found hiding close to and above shorelines. Their odor will then be drawn out across the water and subsequently dissolve into the atmosphere.[54]

Humidity and Rain

It is very difficult for a dog to find game when it is raining hard.[55] Whether the ground surface and vegetation are wet or dry will also influence the dog's tracking success. If the ground surface is damp, the dog's job will be easier.[56] Light rain makes it easier for the dog to follow the trail. Hunting dogs also have much better working conditions when it is foggy or there is water vapor in the air. The air humidity is essential for the maintenance of bacterial activity, while a powerful and long-lasting rainstorm can wash away odors close to the ground.[57] The higher the humidity, the more odorants there are evaporating from the surface and being released up into the air, because they are competing with water molecules for binding sites. That is why forests and marshes have such a strong odor following a rainstorm. Game can control the release of odorants from the glands in their body up to a certain level, but when the air is humid, the volatile compounds from secretions and fur will spread more easily. Just think of how strong the smell of a wet dog is compared to a dry dog. The water molecules occupy the binding sites for the odorants, which makes the smell of the dog stronger because the odorants are in the air.[58]

Turbulence

Turbulence—or eddying—is formed in two ways. Mechanical turbulence is created by the wind and caused by irregularities in the terrain, such as a forest or large stone, while thermal turbulence (convection) is created by rising warm air.[59] It is more difficult for a dog to track game in an area with high turbulence because the width of the scent cone expands rapidly and the concentration of odorants is diffused.[60] In a forest, the trees produce mechanical turbulence. Turbulence will therefore increase when one reaches the edge of a forest. Dogs will consequently have greater difficulties tracking along the edge of a forest than when they are tracking within the forest itself. It is not a coincidence that deer fawns often hide along the edges of forests. Upslope flow is produced in areas that heat up more quickly than their surroundings. On sunny days, logging areas are warmer than the adjacent forest because the rays of the sun heat up the ground, which makes it difficult for dogs to hunt. Logging areas are thus good places for game to hide on sunny days.[61] Upslope flow, such as in the transition from low-lying to high-lying vegetation, will cause the game's scent cone to rise above the dog's sniffing zone, so the dog will not be able to detect the game.[62] The area around large stony surfaces is also very difficult for dogs because these can produce both mechanical turbulence and convection. Stones will also often become warmer than the surroundings when the sun shines because almost all the energy stones receive is converted into heat, while the surrounding surfaces usually contain some moisture, so that some of the energy is consumed by evaporation.[63]

Scent cones can have the same volume, but if there is high turbulence on the sides, the scent cones expand quickly and the odorant concentrations are dispersed across a small area around a bird or other prey. The dog must therefore be quite close to discover the prey. If there is not as much turbulence on the sides, the scent cones will be longer and straighter. If there is more turbulence on the sides, a dog approaching the bird from upwind will detect the bird more easily when the scent cone is longer. The more wind, the narrower the scent cones.[64]

Snow

If there is snowfall before the ground has frozen, the snow will contribute to insulating the ground and keep it warm so dogs can more easily follow a scent trail. If snow falls on frozen ground, few odorants will escape the snow cover, because the ground temperature remains low and dogs will have a harder time following the trail.[65] Cold snow also reduces the bacterial activity.[66] Snow cover disguises the scent of other odorants from life-forms beneath the snow, such as small rodents, making it easier for the dog to follow the odorants of the particular animal it is tracking.[67] Loose and dry snow is very porous, so odorants slip through more easily.[68] In general, snow cover will result in less warmth from the sun being absorbed by the ground, because most of the radiation from the sun is reflected. Both the ground and the air stratum closest to the ground will remain quite cool, and the odorants can be "captured" in this layer of air.[69]

Tracking Down Wounded Game

There is a type of sniffer dog specially trained to track game that has been run over or shot but not killed, leading the hunter to the wounded animal so it can be put down in a humane manner.[70]

If you injure game by hitting them with your car, by law in many places you are required to report this to the local wildlife council. They will provide a sniffer dog, which will follow the trail to determine whether the animal has died or is seriously injured. Deer are unfortunately also wounded every year during hunting season. In the event of wounded game, all hunting is suspended until the animal is found, or until the equipage concludes the search. All members of the hunting team are required to be available during a search. The dog handler is at all times the leading authority during the search and makes the decisions that need to be made.

Tracking Deer Species

In 2011 research scientist Sigbjørn Stokke from the Norwegian Institute for Nature Research led a project that investigated whether specially trained sniffer dogs were able to follow a specific individual animal. Eighteen individual moose and red deer were tagged with GPS transmitters.[71] Prior to the tests, a person frightened a moose or red deer. The animals' last position and the direction of their escape route were recorded. The search equipages tracked the animals two to twenty-two hours later, and they continued for one hour before ending the search. The tracking route of the search equipages was also recorded using GPS transmitters. The research scientists evaluated the equipages' tracking ability by comparing their tracking routes with the trails of the moose and red deer. It turned out that the search equipages were able to follow the trail of the animal they were tracking only to a limited extent, and only one of the thirty-eight equipages was assessed as having carried out an acceptable search. The sniffer dogs often followed the trails of deer species but possibly chose the freshest trail rather than the "cold" trail. The research scientists concluded that this was probably not due to an insufficient tracking ability in the dogs, but rather the dogs' lack of motivation for following "cold" trails. The study was set up in such a way that there was unfortunately no possibility to clarify whether or not the search dogs tracked wounded animals better than healthy animals. It is possible that a wounded animal will increase a dog's willingness to track, so the dog overlooks other trails and follows the injured animal instead. Most people associate such searches with gunshot wounds, and for that reason all specially trained sniffer dogs are trained to follow blood trails.

Normally speaking, a dog will follow an injured and sick animal rather than a healthy one. There are, however, a number of situations where it is preferable to track an animal that is not bleeding. Deer that are shot and hit in a hunting context will usually discharge body fluids, such as blood or abdominal and intestinal juices, but not always. It is neither always the case that an animal that has been hit by a car will have open wounds even if it has been seriously injured. In both cases, it is necessary to have dogs on hand that are able to follow the animal without the need for a blood trail.[72] The specially trained sniffer dogs should receive more training in scent discrimination canine trailing.

Tracking the Brown Bear

The population of brown bears is growing, which will feasibly lead to more cases of wounded animals. The Scandinavian brown bear is not particularly dangerous and as a rule will run away if it is disturbed. In an experiment carried out on GPS-tagged bears in Sweden, master's degree student Gro Kvelprud Moen and colleagues from the Scandinavian Brown Bear Project managed to approach the bears 169 times, and they never experienced any aggressive reaction.[73] However, if a bear is shot and wounded, it can be dangerous.[74] A wounded bear can move at speeds up to 40–60 km/hour (10–15 m/second). It is estimated that almost one-third of all bears that are shot at during the Swedish bear hunting season may be wounded.[75]

Master's degree student Silje Vang and her colleagues from the Scandinavian Brown Bear Project analyzed the tracks of bears tagged with GPS transmitters in 2007 and 2008 in Dalarna and Gävleborg Counties in Sweden. They studied twenty-nine handlers and forty-three dogs and carried out 131 tracking sessions. In 2007 Norwegian and Swedish search equipages tracked bears on average for 214 meters before losing the trail; the successful tracking rate was 49 percent. The training was intensified in 2008 due to the poor results of the preceding year. This produced better results: the Norwegian equipages increased the tracking distance by 34 percent and the Swedish by 29 percent. The equipages with specially trained sniffer dogs and experience with scent discrimination trailing of humans had the best outcomes. This can be an indication that training is necessary to track a specific individual bear. Bear hunting as a training method produced no improvement in tracking outcomes and, in fact, results were to the contrary, with less improvement.[76]

The dogs were able to follow the same individual bear over a period of time only to a small extent. The dogs lost the trail of the bear and switched to other game species in 82 percent of the tracking sessions. There was also a difference between dog breeds. Labradors and Russian hounds were the most accurate when tracking; young dogs performed better than older dogs.[77] A training program has been developed in Sweden with a four-level certification scheme, and Norwegian equipages have used the Swedish method since 2011.[78]

Professor Bjørn Forkman at the University of Copenhagen and his colleagues in Norway and Sweden also did a bear tracking study in

2012 in which twenty-two experienced hunting dogs took part. They found that seven of the equipages did not even start tracking the bear or showed little interest in the trail. Only two of the dogs were able to find the bear.[79] As of 2012, search equipages in Norway and Sweden were not able to track bears in a satisfactory manner.[80]

7

Police Work

The bloodhound Nick Carter, born in 1899, was considered by many people to be the most capable bloodhound in the world. Working with his owner Vonley Mullikin from Kentucky, he sniffed out more than 2,500 finds and suspects, and helped solve 600 cases. On one of the assignments in 1909, Mullikin and Nick Carter followed a trail left by an arsonist that was 105 hours old. The trail started in a field where there was a burned-down henhouse. They followed the trail for 10 kilometers and it led them to a house. A man opened the door and Mullikin said: "Didn't you take sniffer dogs into account when you set fire to the henhouse?" The man replied, "No," and thereby confessed. By asking such a simple question, at a moment when the opposite party was caught off guard and confused, Mullikin was able to make many criminals confess immediately.[1]

In 1888 a series of murders were committed in London by a suspect who was called Jack the Ripper. Police wanted to use two bloodhounds to hunt him down. The bloodhounds Barnaby and Burgho had been trained by

Edwin Brough to track down people. But the dogs were never used in the Jack the Ripper case because the leading dog trainers and veterinarians believed that the dogs would not be able to follow the trail in the slums where the murders had been committed due to the presence of so many other confusing scents. It is reasonable to ask whether bloodhounds trained using the knowledge we now have would have succeeded in sniffing out Jack the Ripper today.

The Main Types of Police Dogs

Dogs have been used to help fight crime as far back in time as ancient Greece.[2] In 1899 police dogs were used for the first time in Europe by the Belgian police force,[3] but it was not until the 1950s that dogs were routinely employed for police work. At this time, German shepherds were usually used.[4]

As of 2014, there are six main types of police dogs used for different work tasks (the categories vary from country to country):[5]

- *Sentry dogs* search for human beings and objects. This can involve missing persons or criminal offenders wanted by the police, such as burglars, car thieves, rapists, and murderers. Sentry dogs are also used in searches for a range of objects such as firearms, ammunition and cartridge shells, or biological evidence such as body tissue, hair, teeth, semen, blood, and dried blood splatter.
- *Cadaver dogs* search for murder and accident victims on land and in water.
- *Narcotics detection dogs* search for narcotics such as cocaine, heroin, amphetamines, and marijuana (see also chapter 8, "Customs and Border Patrol"). The narcotics might be hidden in luggage or vehicles, in the woods or in urban areas (depots), or on a human being. The police also use narcotics detection dogs to search prison cells and in visiting rooms in prisons and schools, at festivals, concerts, and raves, on public transport, and in shopping centers, warehouses, and private homes. In

the United States, the police have recently proposed the use of sniffer dogs to find narcotics plantations/gardens.

- *Explosive and bomb detection dogs* search for different types of explosives and bombs in buildings, sports arenas, schools, luggage, vehicles, ships, and airplanes. These dogs also patrol shopping centers, bus stations, railway stations, and airports (see also chapter 9, "Military"). During the Olympic Games in Vancouver in 2010, dogs were used to search for explosives.[6] Bomb detection dogs are also contracted by private stakeholders in trade and industry to protect office buildings and employees. A bomb detection dog can sniff its way to the person who has planted a bomb, even after the bomb has exploded.[7] These dogs are also used in chemical weapon searches.[8]
- *Search and rescue and rubble search dogs* search for live or dead human beings (see also chapter 5, "Search and Rescue").
- *Arson dogs* are used in cases where there is suspicion of arson. The dogs search for small traces of flammable fluids that are used as accelerants.

In some countries, the police also have dogs trained to search for alcohol, tobacco and cigarettes, foreign currency/banknotes, passports, mobile phones, SIM cards, and agricultural and animal products. Since the turn of the century, customs authorities have used thousands of dogs to sniff out the printer's ink used on banknotes,[9] and they also recognize cotton fibers or banknote paper. Dogs sniff out banknotes of all values and different types of currency. It makes no difference whether it is English pounds, US dollars, or Euros, or whether it is ten- or hundred-dollar bills. Dogs check people who are on their way out of the country by sniffing the person and their luggage. These dogs are trained to react to cotton fibers, printer's ink, and banknote paper. The US dollar is probably the easiest to detect because a more volatile printer's ink is used on this currency that emits a stronger odor. If people have a few banknotes in their pockets, the dogs will not react. However, if they are carrying large amounts of cash on their person or in their luggage (the limit is US$10,000), they will give an indication.

Selection of Police Dogs

Not all dogs have the qualities required to become a police dog and are put through a rigorous selection process carried out by experienced dog trainers/test leaders. In the South African breeding center for police dogs, 70 percent were classified as unfit in 1999.[10] Training a police dog is expensive, and the police naturally do not want to spend a lot of time and resources on dogs ill-suited for such work. It is therefore important to be able to predict at the earliest possible stage how a puppy will turn out as an adult dog. A number of different puppy tests have been developed, in addition to behavioral tests.[11] The best predictors for a potential police dog are its skill in playing fetch when the puppy is eight weeks old and how aggressive the dog is at six to nine months.[12] Potential police dogs can also be given Volhard's puppy aptitude/talent test to determine whether the puppy will become a good working dog.[13]

Scent Lineups to Apprehend Criminal Suspects

Police dogs are trained to recognize people by comparing the scent at a crime scene with the scent of the suspect in a scent lineup.[14] The first person to demonstrate this in connection with police work was Assistant Police Commissioner Bussenius from Braunschweig, Germany, in 1903.[15] The scent lineup works the same way as when a fingerprint in the police database corresponds with one found on an object from a crime scene.

The suspect is asked to hold a stainless steel pipe in his or her hand so their scent is transferred. The dog does not identify the perpetrator but compares the scent on an object found at the crime scene with the scent the suspect leaves on the steel pipe. Such an object can have been stored for up to three years and could be a pistol or a knife or something else that has been used at the scene.[16] Cars, packages, luggage, or envelopes containing money are also common crime scene objects.[17] In Russia and Poland, blood can be used in scent lineups.[18] In a scent lineup using a steel pipe, the pipe is placed at random among other steel pipes bearing the scent of other people who are not suspects in the case, and the dog's task is to recognize the scent of the suspect and bring the steel pipe back to the dog handler.[19] The international standard entails the use of six such steel pipes, of which only one has the right scent, or in control trials where none of them have the right scent. Normally, the test is repeated four times and the steel pipes rearranged at random each time.

Scent identification is done by dogs in a number of countries, including the Czech Republic, Poland, Russia, Hungary, Denmark, and the Netherlands.[20] The results of a range of studies indicate that the dogs' indications are not 100 percent accurate, with the success rate varying from 25 to 93 percent. Sometimes the amount of time that has elapsed since the object was found will influence the dog's outcome.[21] Research scientist Adee Schoon at the University of Leiden in the Netherlands discovered in 2005 that the dogs did everything right when the material was fresh, and the success rate subsequently decreased gradually from week two to six months later (outcomes varied between 25 and 61 percent).[22] Dogs find it easier to identify persons they already know. In spite of significant individual differences, it has been proven that it is relatively easy for dogs to learn how to work on scent lineups. Although the reliability of this identification method is debatable,[23] it is accepted as evidence in courtrooms in Belgium, Poland, Germany, Hungary, and the Netherlands.[24]

The Competition Dog

There are many different kinds of competitions requiring dogs to use their noses. At the Schutzhund World Championship, one of the competitions entails sniffing personal items such as wallets, gloves, pens, scarves, or handkerchiefs and then following the trail of the person who owns the item.

Cadaver Searches on Land

Cadaver dogs were used for the first time by the police in 1974 in New York.[25] The first cadaver dogs in Europe were trained in Austria in 1975 and in Germany in 1978–79. It was not until 1993 that they were introduced in Finland and in 1994 in Sweden. Finding a cadaver can be a huge challenge. Death can be the result of an accident, a suicide, or a murder, and the cadaver might be buried or hidden aboveground. If the cadaver has not been buried, the decomposition of the body takes place much more quickly than when underground.[26] Cadavers can be found in all terrains and environments; cadaver dogs are used to search through buildings, at the scene of accidents (such dogs were used following the Space Shuttle *Columbia* disaster in the southern United States in 2003), and at archaeological excavation sites.[27] In 2011 dogs were used to search for cadavers on large land areas following a conflagration in Texas that burned 1,600 homes to the ground.[28] The cadaver dogs' perhaps most important task is to ensure that next of kin receive information about what has happened to relatives and friends.

Cadaver dogs are trained to detect the scent of human beings who have recently died, human bodies in a putrefactive state, or skeletons, teeth, and blood (more about teeth and blood below).[29] Following death, a number of decomposition processes promptly begin taking place in the body that cause the emission of volatile compounds from the cadaver. The decomposition of a cadaver can be divided into five main stages:

- fresh
- bloat
- active decay
- advanced decay
- dry/remains[30]

Cadaver dogs will also react if the corpse or its fluids have been in contact with the soil or other ground surface and if the cadaver has been in contact with any objects. The amount of time a scent will remain on a given material depends on a range of factors, but we know that the scent in one case lasted for at least one year in a building.[31] The higher the material density, the better the preservation of scent from the cadaver will be, if the period of contact was extensive. For objects with lower material density, such as a mattress, the experience of the police has been that it can be more difficult to achieve an indication, even if the person in question was in physical contact with a mattress for a long time, than for objects with higher material density, such as metal objects.[32] If there has been only brief contact between a human hand and an object, a soft material will quickly pick up more compounds than harder materials. A cotton ball containing a lot of fibers will retain odorants better than polyester. The amount of time a scent remains on an object depends upon the type of fiber and the airstream.[33] If plastic is in contact with an odor source for a long time (weeks or months), it will absorb the scent and retain it for a long time.[34] However, it is not clear exactly what scents dogs detect from a cadaver.[35] Cadaver dogs do not always react to a single odorant, but rather to a number of different compounds produced during the five different stages of decomposition. The dogs should therefore be trained to recognize and react to this entire spectrum of scents.

Often months or even years can pass before dogs are involved in a search: they can find human remains that have been buried for 170 years.[36] Odorants (the volatile compounds) from a cadaver are different from those emitted by living human beings.[37] Four minutes after the heart has stopped beating, the process of decomposition begins and the odor of the cadaver will be produced a mere twenty minutes after the time of death. Milt Statheropoulos, a professor of analytical chemistry at the University National Technical in Athens, Greece, led a study in 2005 that identified more than eighty volatile compounds from two cadavers. The most common compounds were dimethyl disulfide, toluene, hexane, 1,2,3-trimethylbenzene, acetone, and 3-pentanone.[38] Five years later, research scientist Arpad A. Vass at the University of Tennessee found as many as 478 compounds from a cadaver.[39]

It is important that the dog's training can be documented and that several dogs are used (at least three, to compensate for the possibility of errors),[40] because it is then possible to establish with greater certainty that the dogs are giving indications on cadaver scents and not on scents from other things such as blood, rotten meat, or animal cadavers. It is imperative that they be trained using authentic materials, in other words, clothing or scents collected from a dead person. It has been demonstrated that dogs can distinguish between a human corpse and pig cadavers, and that there are large differences in the odorants that are secreted.[41] Dogs can also distinguish between bone fragments from humans and animals.[42] In one study done in 2012, Mary E. Cablk from the Desert Research Institute in Reno, Nevada, and her colleagues compared the volatile compounds from tissue samples (bone, muscle, fat, and skin) from cattle, pigs, and chickens with samples from humans. Although there were common odorants, the animal samples were different from the human samples. If the chemical compounds that were identified are the same as those detected by dogs, the human and chicken samples will probably have a similar scent. The samples from pigs and cows, on the other hand, will most likely not have the same scent as the samples from human beings. The pig samples had just seven of the thirty identified compounds in common with the human samples, and nine of the compounds were only found in the pigs. However, no scientific studies have been carried out confirming that dogs can distinguish between scents from human cadavers and animal cadavers other than pigs.[43] There have been reports of dogs trained solely on the scents of human cadavers that have also given indications for cow cadavers.[44] Dogs trained on pig odors can find human cadavers, and dogs trained on human cadavers can be trained not to react to pigs or other animals.[45] Other dogs will react to the scent of pigs even if they have been trained for human scents.[46] The training is thus of critical importance.

The dogs react to compounds produced by a cadaver such as cadaverine and putrescine, both of which smell like rotten meat.[47] These compounds are often used in imitation cadaver scent products that can be purchased and are claimed to mimic the odor of a corpse. The problem is that these two compounds are found in all biodegradable matter, including human saliva.[48] The chemical composition of two such products was studied in 2012, and it turned out that seven of the compounds had not been previously reported as a product of decomposition. This indicates that these products should not be used in the training of cadaver dogs.[49] A new collection method using a vacuum pump—Scent Transfer Unit (STU-100)—appears to be a better alternative. With the help of this device, the cadaver scent can be collected directly from a corpse and transferred onto cotton balls, which can then be used in the training of dogs.[50] Cotton blends containing polyester and rayon emit more polar compounds than pure cotton. The chemical composition of the material used to collect compounds will determine how many odorants are collected and emitted. The type of material can therefore influence the dog's response and the likelihood of the dog successfully identifying a suspect.[51]

In 2008 some research scientists created a more or less realistic reconstruction of a crime scene.[52] Within three hours after the time of death, they transferred the scents of two dead men, ages sixty and sixty-three, who had been wrapped up in a cotton blanket, to 20 × 20 centimeter swatches of blanket. They did the same thing with scents from people who were still alive. On one of the samples, they spent two minutes transferring the scent, on another, ten minutes. No scent was transferred on some of the fabric swatches. They placed six swatches in glass jars that three dogs were allowed to sniff. In the course of 354 attempts, the dogs did not react to any of the swatches from the individuals who were alive. They reacted correctly to the swatches containing the scents of the dead men in 98 percent of the cases where ten minutes had been spent transferring the scent and in 86 percent of the cases where the scent had been transferred within two minutes.

Underwater Searches for Human Beings

On a cold day in January 1990 in Dyers Quarry, French Creek State Park in Pennsylvania, dog trainer and animal behaviorist consultant

Susan Bulanda and her husband, Larry, were called in to search for two diving instructors who had been reported missing while diving under the ice. The hole in the ice through which the divers had descended was clearly visible. The ice was thick enough to enable them to walk almost everywhere, but in one spot where a waterfall broke the water surface, it was not completely frozen over. As a safety precaution, the Bulandas went out on the ice one at a time. They also worked with a long rope securely tied to a diver. The handlers both had their respective dogs on leads. Scout, a Beauceron, and Ness, a border collie, were the first of their breeds to carry out rescue work in the United States. Because of the temperature, a light fog formed above the water, which caused deer and other wild animals to fall into the water from the steep surrounding mountainsides. Due to the decomposition of the animal cadavers and the vegetation, the temperature was warmest at the bottom of the lake, which can prevent the scent from rising up to the surface. Normally the scent will rise up and pass through porous ice. After Susan and Larry Bulanda made different-sized holes in the ice to enable the scent to rise more easily, it did not take long before the dogs located the area where the two divers had perished. The corpses were found at a depth of approximately 12 meters. The dogs were unable to find the exact location but gave indications close to the site where the cadavers were found. The divers had tied themselves to a rope, but this must have come loose in that it was found around the shoulders of one of the divers. The diver had probably pulled the rope toward him, in the belief that it was securely fastened and then was unable to find his way back to the hole in the ice. Eventually they ran out of oxygen and drowned.[53]

Dogs can also detect the scent of swimmers on the surface or underwater.[54] In 1968 the US Navy started up a water dog defense program for protection from attacks by divers and swimmers on assets such as boats, bridges, and docks. Later the Navy's methods were developed to train search dogs for drowning victims.[55] A problem with drowning cases is that visibility is in many cases extremely poor, making it very difficult for divers to find victims. Dogs can be used to swim out in the water to search and thereby indicate where the victim is located. The dogs swim around in a circle and bark when they find the victim. In conjunction with searches for drowning victims, the police must often search through large areas, and the dog is then usually seated at the stern of the boat. When the drowning victim is discovered, the dog will start to bark.[56]

Dogs can't detect scents underwater the way the American water shrew can.[57] Gas bubbles will rise up to the surface from a cadaver located underwater. The bubbles can come from different fluids or from solid matter such as leather, tissue, bone, feces, or vomit. When the odorants are released into the atmosphere, the dogs detect the scent of the cadavers.[58] Odorants spread four times more slowly in water than in the air. Lipids—fatty substances and wax (ester) and steroids—from the skin glands are not very soluble in water and will produce small odorant bubbles on the water surface.[59]

Water temperature can vary at different depths, which means that the odorants will not rise directly to the surface. They can be transported and come to the surface far away from the source of the odor. The location of the scent will therefore not always correspond with the location of the human body. Also the water current and wind on the surface will have an impact. Dogs can also detect people's scent through ice if it is porous. However, sometimes holes must be drilled in the ice to allow the odorants slip through.[60]

Many people drown annually. When a person has swallowed large quantities of water (wet drowning), the body will sink where the drowning took place and the victim in this case is often found on the bottom. Somebody who jumps off a bridge will usually be found near the place where they hit the water, as long as the search is done within a few hours. This will occur even if the person struggles and the current is strong. If it is a dry drowning, the epiglottis will automatically close to prevent water from pouring down into the lungs. We will usually find people who have died in this way floating on the water surface because the air remaining in the lungs produces buoyancy. After a while the epiglottis will open and the person will then sink.[61] People who have died from dry drowning can be carried far away from the drowning site before the body is found.

The cadaver will remain on the bottom until enough gas has formed in the body cavity to cause it to rise to the surface.[62] If the body floats up, it will normally do so within two to three days in the summertime, while during the winter it can take weeks or months.[63] In other words, it depends on the water temperature. Dogs have found drowning victims months, even years, after they have drowned[64] and have located drowning victims at depths of more than 30 meters.[65] Using dogs enables us to locate drowning victims at a much earlier point in time than would otherwise be possible and perhaps even save lives if we arrive quickly enough at the scene where a person was reported missing.

Searches for Blood, Semen, and Teeth

There are several types of body fluids that can be found at a crime scene or on a victim: vomit, bile, skin oils, blood, semen, saliva, vaginal secretions, and/or urine. It is particularly saliva, blood, semen, hair, teeth, skin abrasions, and nails that are used to identify perpetrators through DNA analysis.[66] In addition to volatile odorants from the hands and saliva, odorants from hair and fingernails can be used to establish people's identities.[67] Body fluids found at a crime scene can also tell us something about what happened: semen suggests that somebody has been raped, and blood that there has been a fight or a murder has been committed. Saliva can be found on a victim, on cigarettes, or on glasses or bottles. Body fluids such as blood, urine, and saliva can also tell us something about whether or not the perpetrator was under the influence of alcohol or narcotics.

Blood and semen in a dried or fresh state constitute important physical evidence often found on or nearby a crime scene. Nonetheless, searching for blood and semen deposits in an area or on an object is relatively new. It is important to find any blood from the victim and blood or semen from the suspect in order to be able to carry out DNA analyses, and thereby provide the police with important evidence in criminal cases. In 2008 the veterinarian Simon G. Newbery from Staffordshire, England, performed an experiment in which a springer spaniel and a border collie were used to search for blood traces on various objects or on the ground. The results showed that an experienced dog was able to detect blood that was from a day to a month old. The dogs detected and brought to the handlers' attention extremely small quantities of blood (0.01 ml) in the grass and on a car seat, and this occurred under different weather conditions.[68] The study showed that dogs can be used to find important biological evidence. In Norway similar experiments have been done with semen and blood.[69]

Teeth are also important evidence in criminal cases, in that they can be used to identify victims. In 2008 research scientist Mary E. Cablk at the Desert Research Institute in Reno, Nevada, carried out a field experiment where dogs searched for teeth with which they had had no previous contact.[70] However, the equipages showed considerable variations—from 20 to 79 percent accuracy, and this was very likely related to the type and amount of training the dogs had received.

Training Norwegian Police Dogs to Find Blood and Semen

From the end of February to the beginning of March 2012, Senior Police Captain Jon Einar Karlsen held a specialized training course at the Police Academy in Kongsvinger, Norway. Equipages from Norway, Sweden, and Denmark would be learning how to search for blood and semen. This was a skill they needed to master to become certified crime scene sniffer dogs. Karlsen put out six cans on a scent detection board: five were without scent and one had the scent of blood or semen. The Norwegian dogs, handled by Kai Iversen (Trixxi) and Harald Grøndahl (Brandfjellet's Kaos Von Kripo, called Kaos), quickly learned how to find the right can. After completing the indoor training, the training was moved outside. A drop of blood was laid out, and Trixxi was given the task of finding this material. It did not take long before the dog gave an indication. Trixxi lay down beside the tree where the drop of blood had been deposited and indicated the find by placing her snout right in front of the drop of blood and one paw on either side of the material so it was not destroyed. Senior Police Captain Per Angel and Sveinung Bakken from Kripos, Norway, could subsequently carry out chemical analyses. First they established that it was blood and not semen, and then secured samples of the blood for DNA analyses.

The Arson Dog

As far back as 1946, bloodhounds were used to sniff out forest fire arsonists. This has contributed to reducing the number of forest fires in several states in the United States where many fires are started.[71] In New Jersey alone, there are 1,600 forest fires every year, and 99 percent of these are caused by arsonists or children playing with matches. In the United States, bloodhounds have been used since 2004 in a variety of different forest fire training programs to raise the awareness of young people in particular, but also of adults. This has reduced the number of forest fires considerably.[72]

In 1984 and 1986, respectively, the initial testing of the arson dogs Mattie and Nellie commenced in the United States.[73] Mattie became the first arson dog, and in May 1987 she had accompanied the police to the sites of forty-one fires. In some cases, her findings led to both an arrest and conviction.[74] In 1995 there were around two hundred active arson dogs in the United States.[75] Star, a black Labrador retriever, was the first arson dog in Europe. Star's owner and handler, divisional officer Clive Gregory, started searching for arsonists in September 1996 through the West Midlands Fire Service in England. Labrador retrievers are often chosen as arson dogs because they are short-haired and have excellent stamina, a calm temperament, and a keen sense of smell.[76]

Searches for Flammable Liquids in Burned-Out Ruins

Arson dogs are specially trained to sniff out evidence in connection with suspicious fires. The firefighters and police officers' task is to try to disclose the cause of the fire: Was it an accident, or did somebody start it intentionally? Fire is, after natural disasters, the cause of the largest losses of both property and human lives. In the United States, arson with the intention of destroying property or taking somebody's life is one of the most common and most difficult crimes to investigate. This is because most of the evidence from the scene has usually been destroyed by the fire. It is said that man's best friend can be the arsonist's worst nightmare.[77]

The dogs' work task is to search through the scene of a fire for small traces of flammable liquids that can be used as an accelerant.[78] These traces are often so small that even instruments cannot detect them, but the dogs are able to find the proverbial needle in a haystack. The dogs can be trained to locate odorants from many different flammable liquids (such as gasoline, natural gas, paraffin, mineral spirits, lighter fluid, brake fluid, solvents, paint thinner, alcohol, and acetone). Dogs can detect this scent for up to eighteen days after a fire was started.[79] They give passive indications for these fluids so the evidence is not destroyed and forensic technicians are able to take samples. This type of indication can help determine where the fire started and if a flammable liquid was deposited where the dog gives its indication. In the event of the suspicion of arson, the remnants from the scene of the fire are sent in for analysis.[80]

In Finland the practice of sending materials in for analysis has been reduced by 80 percent since the use of arson dogs was introduced. The Finnish police have found that the dogs' noses are more effective than laboratory tests. The demand for arson dogs is on the rise, and they are important in solving criminal cases. Insurance companies can also benefit handsomely from the dog's keen sense of smell. There have been cases of private insurance companies that have leased arson dogs to help determine whether or not a fire was set. This can save insurance companies millions of dollars in compensatory damages.[81]

The arson dog's indications are not used as evidence in criminal cases. However, the results of sample analyses can be.[82] The arson dog is also used to search through vehicles because flammable liquids might have been stored there by the perpetrator before the fire was started. Many arsonists like witnessing the fire they have caused and therefore may be standing in the crowd watching the fire. By taking the dog through the crowd, the arsonist can be apprehended. In one case, a dog was able to detect whether the clothes of the suspect have been in contact with a flammable liquid.[83] In Norway it is against the law to search randomly through a crowd. However, if an arson dog on its own initiative should catch a whiff of flammable liquid by air-scent tracking, the dog handler must follow up on this and such a search is thereby legal.[84]

Unfortunately, very little scientific work on the use of arson dogs has been published.[85] In a study carried out by fire inspector Mark Nowlan of Fredericton Fire Department in New Brunswick, Canada, in 2007, dogs reacted correctly in sixteen out of eighteen cases where lighter fluids had been used to start a fire.[86] In a study done in 1997 by chemistry research scientist David J. Tranthim-Fryer at the Chemistry Centre in Perth, Australia, and police consultant John D. DeHaan at the California Department of Justice in Sacramento, it was shown that dogs can have difficulties distinguishing between fire accelerants and burned carpets or plastic flooring.[87] The dog should therefore be trained in a manner that ensures that it is able to make these kinds of distinctions.[88]

The Fire Alarm

> It started silently, late one night in some old electrical wiring. At first there was just a whisper of smoke. Suddenly, the tiny ember burst into flames that quickly spread to the wooden wall. . . . The whole family was asleep upstairs while the deadly fire began to break out downstairs, slowly obstructing the stairway and their way out. In just minutes, escape might be too late. The growing fire and the odorless, toxic smoke would send them deeper into sleep and slowly kill them. . . . It was at this moment that the light-sleeping dog woke up, knowing something was very, very wrong. . . . [S]he barked aggressively at the smoke to frighten it and at the same time alarm her family upstairs, just as she would have if there were a burglar breaking into the house.[89]

This is how the Swedish author and dog psychologist Anders Hallgren's book *Livräddaren på fyra ben* (The four-legged life saver) begins (translated as *Smoke Alarm Training for Your Dog*). Awakened by the dog, the family was able to escape with their newborn baby just in the nick of time. There are stories in the media all the time about how "the dog sounded the fire alarm" and "the inhabitants were able to escape after the dog started howling." Some dogs fortunately react to the smell of smoke, especially if it is thick and rancid. Unfortunately, many dogs die along with their families in fires. In Hallgren's *Smoke Alarm Training for Your Dog*, you can read more about how to train your dog to be a fire and smoke alarm.

The police and the fire department will rush to a potential fire scene in response to reports of suspicious smoke odors. Tarro, Sweden's first smoke detector dog, started working in late 2000. He was also able to detect the scent of gas.[90] The smell of smoke can come from smoldering fires, leaving pots on the stove that can lead to dry-outs, and fires in electrical apparatuses or wiring. However, there is a large difference between the smell of a fire and the smell of smoke from a dry-out.[91] The compound guaiacol emits a typical smoke odor.[92] Determining the location of the source of the odor can be challenging, particularly in large run-down buildings, and without a dog, firefighters will sometimes depart leaving unfinished business at the scene. The dogs, on the other hand, will often find the source quickly, so a large-scale fire can be prevented.[93]

8

Customs and Border Control

When the customs officers boarded trawlers along the coast of Cape Town in South Africa at the start of the new millennium, they were accompanied by Tammy, a border collie—or police agent A1142, as she was also called. In 2000 Tammy was the only dog in the world who could detect the scent of the rare shellfish abalone. Overfishing had put the abalone on the endangered species list, and it was forbidden to export the species to other countries. The Chinese and other Asian cultures consider the abalone a delicacy and an aphrodisiac (a libido-enhancing substance or spice). There is no law forbidding the export of other seafood from South Africa so the smugglers would hide the shellfish with seafood that can be exported legally. As soon as Tammy was on board, she started sniffing around on deck. On one occasion it was not long before she lay down beside a hatchway filled with fish. Hidden among the fish, the customs officers found a shipment of the endangered abalone. Eventually, Tammy was joined by a "bodyguard," a German shepherd named Mac, who also worked for the police. The dogs made a great team.

Tammy found the illegal shellfish, and Mac ran after the smugglers who tried to escape—and tackled them![1]

The Customs Sniffer Dog Worldwide

Many people have certainly encountered dogs at border crossings, on trains, or at airports on their way through customs. The dogs are also used on boats, at seaports, post offices, transport depots, and immigration control points.[2] At a time when the fear of terrorism is escalating, dogs are now being used to search through suitcases for explosives and bombs. They are also used to search through packages sent through the mail.[3] The Belgian sheepdog (Malinois), German shepherd, Labrador retriever, cocker spaniel, and Yorkshire terrier are the breeds most commonly used for this purpose.[4] Even if people are innocent, many become nervous at the mere sight of a dog. But the customs sniffer dog is well trained and will not react just because someone is nervous. The dogs perform a large and varied job as "customs officers."

Travelers carrying illegal products in their luggage must be stopped. Dogs are well-equipped to achieve this objective and are employed at many international airports around the world.[5] The customs officer has the dog on a leash, and the dog sniffs at people and their luggage. It is illegal to bring a variety of food and vegetable products into the United States and Australia. Dogs have been trained to detect these products at airports.[6] The Beagle Brigade in the United States was established in 1984, and these dogs search the luggage of travelers for fruit, plants, and meat that can contain hazardous plant and animal species and that can also have a variety of diseases. The beagles are particularly skilled at distinguishing between different scents and remembering them.[7] At airports they also work at the conveyor belts that transport luggage to and from the airplanes. Here they sniff at each individual suitcase, stepping on them to provoke the release of more odors, before the next suitcase appears on the conveyor belt. They continue working like this until every piece of luggage has been checked. The dogs find shark fins, the bile and gallbladders of bears, live iguanas, snakes, seal penises, parrots, monkeys, and other exotic species.[8]

Poachers unfortunately still hunt for the horns of rhinos and for elephant tusks,[9] which are very popular in China, Vietnam, and Thailand, among other places. It is believed by some people in these countries that these products can have medicinal qualities, which they do not.[10] The horns and tusks are used in Chinese medicine as a cure for cancer and allegedly to drive out the devil. Smugglers can get as much as US$65,000 for one kilo of rhino horn, which means that such horns are more valuable than gold.[11] In Gabon more than 11,000 elephants have been slaughtered by poachers since 2004, and the population has been reduced by 30 percent.[12]

TRAFFIC India

On July 29, 2010, the forest management authorities of the three states Madhya Pradesh, Maharashtra, and Jharkhand in India acquired five new employees: dogs that in the course of three months had been trained to detect the scent of illegal game products. They sniff out both the bones and skins of tigers and leopards, along with bear bile—products that are used in traditional Chinese medicine. The organizations TRAFFIC India and WWF-India have funded the training of these dogs. They have also established sniffer dog programs targeting smuggled game products in several other countries, including Russia.

A few months after completing the training, the German shepherd Raja helped to find a leopard killed illegally that poachers had attempted to hide. The dog also played a part in tracking down the culprits. Another dog, Jackie, tracked down two poachers before they were able to do any damage. The dog Tracey assisted in the detection of two elephant tusks at the Dalma Wildlife Sanctuary in Jharkhand. When the forest management authorities found the dead elephant and saw that the tusks were missing, they initiated a search. They were unable to find the tusks. Tracey was therefore called in from the Palamau Tiger Reserve in the Betla National Park, and after a meticulous search she found the hidden tusks. They weighed more than 32 kilos. So far, the program has been a success, and TRAFFIC India and WWF-India hope that more states in the country will begin using dogs in conjunction with their work.[13]

Searches for Narcotics

In the 1970s, the US customs service began putting dogs to work in the fight against drugs.[14] One of the reasons people started training dogs was that soldiers tried to smuggle marijuana into the United States during the Vietnam War in 1971.[15] The dogs could detect the scent of marijuana even if it was stored in glass jars or sealed plastic bags doused with perfume and/or spices or if it was mixed with other substances. This success led to an expansion of the program, and the dogs were later trained to detect harder narcotics, such as cocaine and heroin.[16] Twelve narcotics dogs that worked in the border city of El Paso, Texas, found in the course of nine months narcotics worth US$100 million.[17] Smugglers go to great lengths to hide the scent of illegal narcotics, so the dogs must receive a lot of training. Some dogs become so proficient that smugglers put bounties on their heads.[18] It is interesting to note that in the United States, cocaine residue is found on at least one-third of the money in circulation.[19]

At border crossings, vehicles are often thoroughly checked. If narcotics are hidden in a vehicle, the scent can slip out through openings if there is a wind blowing from the other side, so sniffer dogs are able to detect the scent of narcotics quickly and from considerable distances.[20] They will also sniff the vehicle thoroughly, both inside and out. The dogs can find narcotics even if they are floating in the fuel tank of a vehicle, and even if a package of drugs has been welded into a metal container. It takes the dogs five to ten minutes to search through a car. A manual check done by a customs officer can take at least twenty minutes[21] and in some cases up to three hours.[22]

It is not always easy for narcotics dogs to find the source of the drugs. One dog went completely "crazy" inside a container full of Oriental furniture. There was a very strong odor inside the container, but the dog was unable to locate the source of the scent. It turned out that the furniture had been varnished with cannabis/resin, so the entire container was the source.[23]

Tadeusz Jezierski, professor of behavioral biology at the Department of Animal Behaviour in Jastrzebiec, Poland, and his colleagues demonstrated in 2014 that dogs found narcotics in 83 percent of the cases when it was hidden inside a familiar or unfamiliar room, but they were not as effective when they searched outdoors (64 percent) or in cars (58 percent). Marijuana was the narcotic substance that the dogs had the easiest time finding. They had the greatest difficulties finding heroin, followed by cocaine, amphetamines, and hash. Hash was identified easily even forty-eight hours after it had been removed from its location in a room. The dogs were not particularly adept at detecting the scent of heroin after forty-eight hours. The German shepherd was the dog breed with the best outcomes, followed by the English cocker spaniel and the Labrador retriever, while terriers were the worst.[24]

Searches for Mobile Telephones

Many things are smuggled into prisons. It is not only a matter of drugs, but also mobile phones. For example, when mafia leaders have telephones in prison, they can continue running their criminal activities from inside. Although the telephones can be hidden anywhere in a prison cell, the dogs sniff out the gases emitted by the telephone batteries.[25]

Searches for DVDs and CDs

Pepper, Lucky, and Flo were the first dogs trained to discover bootleg DVDs in the 1970s in the United States. They detected the chemical compound (polycarbonate) used in DVDs and CDs.[26] They have sniffed out millions of illegal DVDs, CDs, and video games. The Motion Pictures Association of America lost billions of dollars every year because of bootleg DVDs. Lucky and Flo worked in 2007 in Malaysia's capital, Kuala Lumpur. Here they found a secret room in a video store where 150,000 illegal DVDs were hidden. A month before this, they had found a million illegal DVDs and CDs, valuing almost US$3 million. After this find, the bootleg film perpetrators offered up a reward of US$30,000 for anyone who managed to shoot the dogs.[27]

9

Military

Cairo is a specially trained military dog who works for the Pentagon in Abbottabad, Pakistan. He can sniff a piece of clothing and then find the person it belongs to, even if the scent is several days old. He can lead soldiers to a closed-off room where a person being searched for is hiding. He can detect the scent of explosives, tripwires, and other booby traps. It is said that on the night of May 2, 2011, he was lowered from a helicopter along with members of a US Navy SEAL team. Cairo was searching for Geronimo, which was the code name for Osama bin Laden, the world's most wanted terrorist. An infrared camera was strapped to Cairo's back, and it picked up all the activity in front of him. He also had tiny earbuds in his ears that were hooked up to a wireless transmitter. Commands could then be whispered to Cairo from several hundred meters away, while it was also possible to see what was happening as he moved forward. Geronimo was killed in an exchange of gunfire that night, and Cairo had taken part in yet another successful mission. Dogs also played an important part in the search for Iraq's former leader Saddam Hussein and his two sons.[1]

The Contribution of Dogs in Wartime

Throughout all of history, dogs have been used in the context of war.[2] The Romans and Napoleon used dogs in warfare.[3] During World War I, the Belgians, the French, and the Germans all used dogs to find wounded soldiers.[4] The Airedale terrier breed was frequently used during World War I to carry messages and medical supplies to soldiers behind enemy lines. If a soldier was found who was unconscious, the dog would return to its handler and lead him back to the soldier.[5] Medical service dogs tracked down and alerted medical staff about wounded and dead soldiers stretcher bearers had been unable to find because of fog, fire, darkness, or other difficult conditions.[6] The British used dogs to locate missing persons in bombed building rubble during World War II.[7] Different races of human beings have different odors, and dogs were trained to find Japanese soldiers during World War II;[8] scout dogs would sound the alarm when they detected the scent of hostile individuals or enemy troops nearby.

During the Algerian War in 1954–62, sniffer dogs were also used to find enemy troops who had escaped.[9] And during the United States' involvement in the Vietnam War (1965–73), more than 4,000 American military working dogs were used, and it is estimated that they saved more than 10,000 human lives.[10] The dogs were especially effective at night. The US military used ally South Vietnamese soldiers to create trails so the dogs could be trained to detect the scent typical of the enemy as well as locating North Vietnamese billeting areas and tunnels. In the attempt to fool the American military, the North Vietnamese washed with American military soap and covered the tunnel vents with T-shirts from American soldiers.[11] They also spread sap from a plant on their bodies to try to reduce the chances of being found by the dogs.[12] Dogs were also used to find drowned pilots and to track down North Vietnamese who would use reeds to breathe underwater while trying to infiltrate the Americans' water route. The dogs could smell the swimmers one and a half kilometers away and up to depths of nine meters. The North Vietnamese did what they could to kill the dogs and would receive a reward if they managed to bring back an ear tattooed with the dog's ID number.[13]

Messenger dogs carrying important information or ammunition could travel distances of five to ten kilometers. At the end of such a journey, the person waiting for the dog would have to be someone the dog knew well. The dog also had to be familiar with the route in advance or a tracking fluid was used with a characteristic scent that the dog had been trained to follow. The enemy tried to confuse the dogs by laying out fake trails, so the special fluid had to be top secret.[14]

Dogs have subsequently been used to serve in wars in countries such as Iraq, Afghanistan, and Pakistan. Many people have been saved by these hardworking dogs. Strapped to their handlers, they can be dropped by parachute with oxygen masks, cameras, and other advanced equipment attached to their bodies. In 2011 the Pentagon alone had 3,000 work dogs prepared for duty in the armed forces all over the world.[15]

The poodle, St. Bernard, and pointer were the first dog breeds to be used in warfare. German shepherds have been the most commonly used military working dog breed, but use of the Belgian sheepdog (Malinois) has steadily increased since 2000. As of 2011, the Dutch shepherd dog, the Doberman pinscher, rottweiler, Labrador retriever, and springer spaniel have also been used in the context of warfare.[16]

Searches for Mines and Booby Traps in Rural Bosnia

Lieutenant Asbjørn Grande has taken part in many dangerous international operations. Once in the spring of 1998, he was going to search for mines and improvised explosive devices (IEDs) out in the Bosnian countryside. The military mine detection dogs search through unfamiliar minefields at night, and their handlers would wear night-vision goggles with infrared lights. Civilian humanitarian mine clearing is done using more controlled methods during the daytime and under the right weather conditions. The searches are done only after a mine flail has first driven over the area and detonated land mines in its path. The military mine searches are extremely dangerous missions. In the area where Grande was supposed to search, there had been very intense fighting, so the mine density was high and many houses were rigged with IEDs. Grande had brought along Serbian map sketches of the minefield, mine groups (more than nine mines), and buildings containing IEDs. He had also been on a mission in Bosnia the preceding autumn and cleared away many mines, but now he was in an area with a lot of nettles where it was difficult to acquire an overview. Many of the antipersonnel mines had been laid strategically, something he was able to discern from the Serbian mine map sketches. According to the sketches, some of the mines here were small antipersonnel mines that were rubberized on top and contained the explosive compound tetryl. He also knew that the Croatian population did not want the mines and IEDs to be removed from the Serbian part of the area, because this would allow the Serbians to move back. Grande therefore had a strong suspicion that somebody within the Croatian community might set new booby traps.

Grande worked on the field with the mine detection dog Lisa and a mine clearer. They marked out a 40-centimeter-wide path and began searching along the sides of a house. Around this house there was a type of concrete platform that was one meter wide. According to the sketch, there was a mine against one corner of the house. Grande also saw there was a pile of dead grass lying there. As they started to search at the opposite corner, Lisa didn't seem to be making much progress. The dog stopped several times. Any scent here should be easy for her to detect so Grande was a bit puzzled about why she kept stopping. They approached the corner of the house. When they were only two meters away, Grande saw that Lisa lifted her muzzle. He corrected her with the words "Lisa search." Grande understood then that something was happening. Lisa put her nose on the grass pile and stopped. She calmly stuck her snout down into the grass. Then she pulled back her head and gave a clear indication by sitting with her head turned to face the hole her snout had just made. Grande made Lisa back up one meter, so she was between his legs, and he bent down to look. It felt as if his heart stopped. Grande said to the man behind him: "Stand completely still, don't move one inch." They were standing on a tripwire attached to a stake mine under the pile of grass. There was not supposed to be a stake mine or tripwire here. Somebody had taken away the mine that was supposed to be there and installed a stake mine instead and put in a tripwire underneath the moss—a diabolical trap. Had Lisa taken one more step and put her paw on top of this wire, their lives would have come to an abrupt end. Grande froze for a few seconds while thinking, and then he fished the cutting pliers out of his vest pocket. He bent down and cut the wire directly up against the mine. Lisa sat completely still as she is supposed to do in this kind of situation. It was the tripwire that had confused Lisa on the way in. Grande picked Lisa up and turned her around, and they walked out the way they had come in. When they got to safety, they breathed a sigh of relief. Grande was drenched in sweat. The margins had been in their favor this time as well. They could thank Lisa for that. She did exactly what she had been trained to do.

Explosives and Bomb Detection Dogs

Explosives are made of a mixture of different chemicals that explode. Some compounds are more prevalent than others.[17] There are many different types of explosives, and each group has a unique chemical composition.[18] Dogs use only their sense of smell and not their sense of sight when searching for explosives.[19] They can be trained to find explosives such as dynamite, Semtex (plastic explosives), FORMEX, hexogen (RDX), C-4, nitroglycerin, ammonium nitrate, plastic explosives, and smokeless propellant gases and fuses.[20] One of the most common high explosives of the last one hundred years is 2,4,6-trinitrotoluene (TNT). TNT not only constitutes a security threat but is also a threat to the environment.[21] Tadeusz Jezierski and his colleagues demonstrated in 2012 that dogs had the hardest time detecting TNT, followed by explosives, gunpowder, Semtex, and dynamite.[22]

The "holy grail" is determination of the particular scent that dogs will use to detect these explosives. The most practical solution is to train the dogs on the volatile odorants found in the explosive. The problem with this is that different explosives can be made of different substances.

Different manufacturers use different materials, which makes it important to train dogs using a broad and varied range of materials.[23] Dinitrotoluene (DNT) is also produced as a by-product and is a key substance in TNT detection, while 2-ethyl-1-hexanol is the key substance for RDX—two key substances that dogs should be trained to detect.[24] It is important to remember that the scent of small concentrations of explosives can differ from the scent of large concentrations.[25] In other cases, the amount of the material—for example, if there is 1 milliliter or 10 milliliters of the substance nitromethane—was of no significance to the success of the dogs' outcomes (54 percent in both cases).[26] If dogs are trained on individual substances, they may have difficulties detecting this substance when it is found in an unfamiliar mixture.[27] If you are going to train a dog to find C-4, it is recommended that you train it on the explosive in question and not on imitations or individual substances, as has formerly been the practice. Some of the volatile substances in explosives are also found in ordinary products such as PVC tiles, PVC pipes, insulation tape, and credit cards.[28]

Bomb detection dogs are used by both the armed forces and the police. They are most commonly from the breeds German shepherd, Belgian sheepdog (Malinois), Labrador retriever, springer spaniel, and vizsla.[29] These dogs normally remain calm both in large crowds and with strangers. One of the world's first bomb detection dogs was a German shepherd named Brandy. Immediately following the departure of a plane traveling from New York to Los Angeles on March 9, 1972, word came that a bomb had been planted onboard. The plane turned around and it was only by chance that Brandy happened to be at the airport that day. Right after the passengers disembarked, Brandy was taken on board. The dog searched through the plane, and when she reached the cockpit, she gave an indication for a suitcase. There was a bomb in it, and luckily the team was able to defuse it twelve minutes before it was supposed to go off. From that day on, dogs acquired an important role in bomb detection work at major US airports. It was President Richard Nixon who initiated this program so travelers would feel safer at airports.[30]

The dogs also detect bombs and other explosives if people have such things concealed on their person when they walk through an airport or railway station.[31] These dogs can detect the scent of a person carrying a concealed bomb even fifteen minutes after the person has left the area. Amtrak began using these types of dogs on its trains and in its railway stations in 2008, and they were implemented at the Los Angeles International Airport in 2011.

The Explosives and Bomb Detection Dogs Sadie and Buster

A few meters away from the parking lot, a bomb had been buried under a pile of sandbags. A German soldier had just been killed by a car bomb, and there was a feeling of nervousness in the air. Planting an additional bomb is in fact a preferred terrorist tactic, designed to do the greatest possible damage. If this bomb also went off on this November day in 2005, another two hundred UN soldiers and rescue crew members in Kabul in Afghanistan would have died or been seriously injured. The NATO soldiers were fortunately not alone in Kabul on this day. The group had with them Corporal Karen Yardley from Scotland and Sadie, a black Labrador retriever. The odorants from the bomb had reached Sadie's nose, and her black tail was wagging energetically. She sat down and stared straight into a brick wall. Corporal Yardley recognized Sadie's signal and understood that there was a bomb on the other side. Yardley immediately notified the bomb disposal squad, who subsequently successfully deactivated the bomb. The soldiers' lives were saved thanks to Sadie's sensitive nose.[32]

The springer spaniel Buster is another well-known military dog.

> "The soldiers had found nothing so I unleashed Buster and sent him in," said the hound's handler, Sergeant Danny Morgan of the Royal Army Veterinary Corps. "Within minutes he became excited in a particular area and I knew he'd discovered something. The Iraqis we spoke to had denied having any weapons. But Buster found their arms even though they'd hidden them in a wall cavity, covered it with a sheet of tin then pushed a wardrobe in front of it."

> Buster, who is thought to be the only explosives sniffer dog working with the coalition in Iraq, has been given his own protective gear in case of chemical attack. When gas or missile attack warnings sound, he leaps into a special sealed pen equipped with an electric motor that pumps air through a gasmask filter.
>
> "I trained him by teaching him to fetch weapons like guns and ammunition instead of sticks and balls," said Morgan, who keeps Buster as a family pet in Hampshire, England. "He loves his job simply because he thinks it's a game and obviously has no idea he's going into dangerous situations."[33]

The Dickin Medal is awarded to animals to honor service with distinction in wartime. Both Buster and Sadie received this medal for their heroic contributions during these missions.[34]

Searches for Mines and Unexploded Munitions

The most valuable work that dogs carry out is very likely the clearing of land mines and other unexploded munitions from areas of land after wars.[35] The first mine detection dogs were put to work during the Second World War. Dogs were also used during the Korean War and the Vietnam War. The Swedish military began using mine detection dogs in the 1950s.[36] Mine detection dogs have thus been in use for more than seventy years, but dogs have been used for humanitarian demining since 1989. When the Russians pulled out of Afghanistan in 1989, mine detection dogs came to play an important role.[37] As of 2013, they were found in almost every country that has a problem with mines. Usually German shepherds, Belgian sheepdogs (Malinois), or Labrador retrievers are used.[38]

The Geneva International Centre for Humanitarian Demining (GICHD) is an international organization based in Switzerland that works to eliminate mines and other dangerous explosives.[39] GICHD was founded in 1998 and has fifty members from many different countries. Their work is supported by many governments and organizations, and they visit some sixty countries a year.[40] Another important organization is Norwegian People's Aid, which has its own training center for mine detection dogs (Global Training Centre), located outside of Sarajevo in Bosnia and Herzegovina. This training center is one of the best in the world. Terje Groth Berntsen is the director of the center, where they breed dogs and carry out training of dog handlers and dog trainers. The dogs are used in the center's own demining programs and also by other demining operators. They have worked with different breeds, but as of 2014 use only Belgian sheepdogs (Malinois). A Malinois is able to clear 30 percent more of a minefield than, for example, a German shepherd in hot weather conditions. Along with being lighter and smaller than German shepherds, Belgian shepherds have a coat that keeps them cooler, which means less panting. They also have an extremely strong willingness to search. The dogs receive training for a year and a half and can work for up to ten years. There are strict requirements imposed on a mine detection dog. To be a good mine detection dog, the dog must have the utmost desire to search, since they must be in the field from four to five hours at a time, in 30–40°C heat, every day. In other words, such a dog is like an elite athlete. The Global Training Centre has been dispatching dogs to different minefields since 2004.

As of the writing of this book, the mine detection dogs trained by Norwegian People's Aid have demined an area equivalent to almost 79 million square meters in different countries all over the world. No mines have subsequently been found in these areas. Since 2006 the dogs have found 2,896 mines, 8,850 explosive remnants of war (such as unexploded shells and ammunition), and 278,433 different fragments of a variety of explosives. No accidents have occurred so far, and no dogs have been injured or killed by mines or other explosive ammunition.[41] If the soil contains a lot of stones and metals, a deminer with its instruments will take much longer than a dog. The dogs find the mines on average twenty times more quickly than a deminer. In the course of a working day, 800–1,700 m^2 of cultivated land can be demined using dogs,[42] while a deminer will only manage around 10 m^2. In 2006 Norwegian People's Aid had sixty-five active mine detection dogs working in five countries, and it has trained many more subsequent to this, approximately seventy-five per year. The dogs are sent to countries such as Cambodia, Ethiopia, Jordan, Republic of the Congo, and Tanzania. In Cambodia the dogs have found mines buried six meters underground, even though they have been lying there for forty years.[43]

The dogs sniff out the scent of TNT, which is found in traditional metal land mines and also in newer mines covered with plastic. TNT is found in approximately 80 percent of all mines.[44] TNT compounds that leak out of the mines slowly creep up to the surface and are bound to dust particles on it.[45] The compounds from the mines creep more slowly to the surface when in dense clay than in loose sand. The dogs can detect the scent of the compound even in very small quantities.[46] The odorants from the mines are easier to detect if the dogs sniff close to the ground.[47]

When peace is declared, the mines are still hidden in the ground. Children out playing can step on the mines and be killed or seriously injured. Every year more than 15,000 mine accidents occur in more than eighty-two countries. It is not only human beings who are killed by mines. Between 1994 and 2005, approximately 300,000 wild and domesticated animals were killed by mines.[48] There are more than 100 million mines throughout the world today. The possibility of a mine in a given area can mean that entire villages cannot return home or cultivate the land.[49] It is estimated that it costs US$5 to put a mine in the ground, while it costs about US$1,000 to remove it. In 2002 mine detection dogs were used in twenty-three countries, and more than seven hundred dogs were on the job.[50]

Most dogs are used in the field, where they search for mines directly. In the context of this work, it is important that the dogs give 100 percent correct indications when there are mines, but they can make mistakes (5 percent) if no mines are to be found (in other words, give indications for mines even when there are none). Dogs are trained first in a test minefield where mines without a trigger mechanism have been planted.[51] After fifteen weeks of training, the dogs achieve outcomes of approximately 95 percent correct indications. In a field search, more dogs are used in order to compensate for the errors they may make.[52] If a mine detection dog makes a mistake, in the worst case both the dog and the handler can be killed. The dogs are taught to be calm when they give an indication for detection of a mine in order to avoid detonating the mine. A single jump or other movement can kill or injure the equipage. The dog handler informs the demining team by walkie-talkie and marks the detection on a map. The mine is deactivated before it is moved to a secure area, where it is detonated.[53]

The dogs do not always work out in the field. They can also receive the scent on a filter gathered from a minefield. This is called Remote Explosive Scent Tracing (REST).[54] This technique was first implemented by the demining company Denel Mechem in Mozambique and in Angola in the early 1990s.[55] This effective method has been used in a number of countries that have mines for the quick clearing of roads.[56] In practical terms, the demining team uses a vacuum pump to suck the air over a road into a filter.[57] Typically, searches of a stretch of road 100 to 200 meters long and approximately 5 meters wide will be carried out at a time. Different sections of the road are gradually mapped out, so many filters are collected. The filters are taken to the laboratory, where they are presented to trained mine detection dogs on a carousel or a scent detection board. The dogs then find the sample that has the scent of a mine.[58]

10

Medical Detection

Rebecca Farrar from England is seven years old and has type 1 diabetes. She has to test her blood sugar seven times a day. She must consume glucose if her blood sugar is too low and take insulin shots if her blood sugar is too high; if not, she will suffer a physical collapse. Luckily, in 2010 she acquired a dog that reacts when her blood sugar level drops or rises. Shirley, a Labrador-golden retriever, goes to school with Rebecca and was the first canine assistant allowed into the classroom in Great Britain. Shirley sits beside her desk in the classroom along with twenty-eight other children. By the time Rebecca experiences a dramatic drop in her blood sugar level, without having demonstrated any other symptoms, the dog has already warned the teachers. If Rebecca's blood sugar is not as it should be, Shirley lets her know by licking her hand. She even makes sure that Rebecca has on hand a box of supplies so she can check her blood sugar. At night the two of them sleep side by side. Shirley has saved Rebecca's life many times. A serious episode could cause Rebecca to fall into a diabetic coma, and she could die if the symptoms are not treated immediately.[1] Another dog by the name of Mr. Darcy had his own Facebook page, where it was possible to follow along with the work he did for his owner, Abi Atkinson. He also reacts to high and low blood sugar levels and takes care of her every single day.

The Health Benefits of Having a Dog and the Odors of Different Illnesses

The dog is often referred to as man's best friend. There is a reason for this: dogs can reduce stress levels, anxiety, and depression and, beyond these benefits, ensure that their owners have company, entertainment, and get physical exercise. There is a lot of evidence demonstrating that people's health can be improved by having a dog[2] and that they can also help reduce the risk of many less serious illnesses/ailments, such as headaches, the common cold, fever, and dizziness, as well as more serious cardiovascular diseases, such as blood clots and heart attacks. People who have a dog smile more and are greeted by and fall into conversation with other people more frequently.[3] There are more benefits to owning a dog if you are single than if you are married, and more benefits if you are a woman than if you are a man. However, some married couples may find that it leads to a lot of additional responsibility if they are also juggling parenting duties, their jobs, housework, family, friends, and leisure activities.[4]

In recent years, we have become more aware of dogs' ability to serve as an "early warning system" for certain kinds of illnesses.[5] We know that medical personnel can associate different odors with different illnesses, and if our urine or feces have an abnormal odor, this can help alert us to a health-related issue.[6] Many illnesses emit specific odors and can to a certain extent be identified on the basis of this.[7] In the past, doctors would diagnose illnesses by studying patients' body fluids, urine, and feces.[8] They knew that typhus smelled like freshly baked brown bread, measles like recently plucked feathers, and yellow fever like a butcher shop.[9] Some skin diseases also have characteristic odors.[10] Scurvy will give you an odor that smells rotten, and mononucleosis has a sourish odor.[11] If you have a uremia, your urine will have an unpleasant ammonia scent, and even schizophrenia produces a characteristic odor (from the substance trans-3-methyl-2-hexenoic acid, which is found in sweat).[12] It can be difficult for humans to recognize the odor of different illnesses, since variations between individuals can cause the production of different odors from the same illness.[13] For that reason, it's not strange that dogs have been trained to detect different illnesses using their sense of smell.

Dogs Can Detect Diabetes Type 1 and 2

According to the International Diabetes Federation, 366 million people had diabetes in 2011, and the number will very likely increase to 552 million by 2030.[14] Type 1 diabetes is a chronic illness caused by a deficiency of the hormone insulin and is the most common metabolic illness found in human beings. People who have type 1 diabetes must take insulin in the form of injections. High blood sugar can develop into acidosis (ketoacidosis) with type 1 diabetes.[15] In the urine, ketoacids are converted into acetone, so the breath will smell of acetone.[16] Type 2 diabetes is a chronic illness caused by a deficiency in the hormone insulin and/or diminished insulin resistance. Eventually the metabolic changes will affect all body functions, causing lethargy, fatigue, and low spirits.[17]

Hypoglycemia is the medical term for a condition in which the blood sugar level is below normal, while hyperglycemia means that the blood sugar level is higher than usual. Low blood sugar is most common in people who have type 1 diabetes and take insulin, but it also arises with the use of insulin for type 2 diabetes.[18] A low blood sugar level causes reduced brain function, and both the ability to think and reaction time are diminished. People who have this illness are at greatest risk just before meals and at night.[19] A big problem is that these people are often unable to recognize by themselves that their blood sugar level is dropping. It is unclear how dogs are able to detect hypoglycemia. Odor has been proposed as the most probable explanation,[20] although in an experiment carried out by research scientist Ky Dehlinger at the Medical Center in Portland, Oregon, and colleagues in 2013, three dogs were unable to recognize the skin odor from the arms of people with type 1 diabetes.[21] It has also been suggested that there can be a connection with the vomeronasal organ (see chapter 2, "A Dog's Sense of Smell").[22] Increased sweating has been reported repeatedly in people suffering from hypoglycemia.[23] It is therefore probable that dogs detect the chemical change in their owners' sweat.

Many dog owners who have diabetes have reported that their dog will nudge them until they wake up, others that their dog wakes them up by barking and scratching on the door when their blood sugar level drops.[24] It is very likely that these individuals did not send any signals other than an odor. The dogs did not calm down until the owner had eaten something and the blood sugar level returned to normal. An elderly farmer who had type 1 diabetes had his dog with him in the car one day when he was out driving in France in 2006. When the owner's blood sugar level dropped, the dog woke up, stared at the owner, and started barking furiously until he stopped the car.[25]

It wasn't until 2008 that research scientist Deborah Wells and colleagues at Queen's University in Belfast carried out a study on the reactions of untrained dogs to hypoglycemia in patients with type 1 diabetes.[26] They found that many dogs (65 percent) detected hypoglycemia before their owners were aware that they were experiencing symptoms. The dogs reacted by barking, licking, rubbing their noses against the owner, jumping up on them, or staring intensely at their face. Some dogs reacted with fear. They ran away, hyperventilated, or started to shake. The dogs also woke their owners up during the night. In the majority of the cases, the dog was sleeping in another room and if the door to the bedroom was shut, the dog scratched on it to wake the owner. In 2013 animal welfare and behavioral biologist Nicola Rooney at the University of Bristol in England and colleagues trained dogs to respond to people with hypoglycemia and hyperglycemia. The dogs were greatly appreciated by their owners, increasing their quality of life.[27]

Kiko "the Surgeon"

There are many incredible dog stories in circulation. High on the list of these is the story about the Jack Russell terrier Kiko, who became a "surgeon" by amputating half the big toe of his owner Jerry Douthett from Rockford, Michigan. He basically just bit off the toe and ate it. Both Kiko and other dogs had been showing a great interest in the big toe for a long time. They were always sniffing at it. One evening, after the owner had had a little too much to drink and was in bed asleep, Kiko took advantage of the occasion. The owner woke up with a big toe that was nothing more than a bloody stump. At the hospital, the doctors discovered that the owner had type 2 diabetes and a serious infection in his big toe. His blood pressure was dangerously high and the doctors had to complete the amputation.[28]

The Diabetes Dog Nemi

One diabetes dog named Nemi assists seven-year-old Nicklas André. Nemi—a German shepherd, rottweiler, and Labrador retriever mixed breed—reacts when Nicklas André's blood sugar level drops and also warns his mother, Anneli G. Johansen.[29] If his blood sugar level is below 4.2, Nicklas André becomes hypoglycemic, his body debilitated and his speech lethargic and slurred. During the last months before Nicklas André learned that he had type 1 diabetes, Nemi was very restless. When the family came home after he had been in the hospital, Nemi sniffed at Nicklas André a little before settling down on the couch. The dog was "suddenly" calm. She now gives a warning long before the other family members notice anything, even if Nicklas André is lying in bed asleep upstairs. Nemi had not had any training; she reacted instinctively. Later Johansen began training Nemi using saliva samples on a cotton ball to develop her innate abilities even more. She found a dog training school in Virginia, Tidewater K-9 Academy, that could help her train Nemi to be a diabetes dog. (A number of books and articles have also been written about the training and selection of diabetes dogs.[30]) More effective training commenced, and Nemi had a powerful reaction when Nicklas André's blood sugar level rose to around 15. Teaching the dog to respond to low blood sugar levels was a more difficult task, but Johansen managed that as well. In the freezer she keeps bags of odor samples that she collected when Nicklas André's blood sugar levels were low and high to train Nemi to learn to recognize.[31]

Dogs Can Detect Cancer

In 2008 close to 12.7 million people were diagnosed with cancer worldwide, and 7.6 million people died from the disease.[32] Two years later, in 2010, as many as 28,271 people developed cancer in Norway. The most common forms for men are prostate and lung cancer, while for women the most common are breast and colon cancer.[33] Research has shown that the early detection of cancer is very important, and increasingly more dogs are being trained for this purpose. It all started in 1989 when a forty-four-year-old woman had visited her doctor at King's College Hospital in London after her mixed-breed dog (Doberman/border collie) had started sniffing at one particular mole on her left leg. The dog showed no interest in the other moles. The dog sniffed at the mole several minutes every day, also through her trousers, and this behavior continued for several months. The dog had even tried to bite off the mole. It turned out that the dog had detected cancer in the mole.[34] Twelve years later, Parker, a Labrador retriever, detected cancer on the thigh of a sixty-six-year-old man from Nottingham, England,[35] and in 2005 a dachshund puppy discovered breast cancer in a forty-four-year-old woman from Wausau, Wisconsin.[36] In both cases, the dogs lost interest in sniffing these parts of the body after the cancerous tumors were removed.

Cancerous tumors produce odorants (alkanes and benzene derivatives) that are emitted into the air through people's lungs, urine, feces, tissues, blood, and sweat.[37] Dogs can detect these compounds even in extremely small concentrations. Some dogs possibly have an innate ability to detect the odor of cancerous tumors, while others can be trained. One possibility is that the volatile organic substances and compounds produced by the tumors and emitted through the sweat, breath, and/or urine stem from the group of genes called the major histocompatibility complex (MHC).[38]

Since 2004 a number of different scientific studies have shown that dogs can detect bladder cancer, melanoma, lung cancer, breast cancer, ovarian cancer, prostate cancer, and colorectal cancer.[39] The research scientist Carolyn Willis and colleagues at Amersham Hospital in Buckinghamshire, England, performed the first scientific study in 2004 designed to investigate whether dogs could detect the odor of cancer.[40] They began by trying to train dogs to detect the odor of bladder cancer in patients' urine samples. The research scientists hypothesized that substances from the cancer cells could be found in the urine, and that it is these substances that give the urine a characteristic odor. Formaldehyde has been found in urine samples from patients with bladder and prostate cancer in much higher concentrations than in patients who do not have cancer.[41] It took the research scientists seven months to teach the dogs to recognize the odor of bladder cancer. The dogs were trained to ignore the urine from people who smoked. The dogs also had to learn to distinguish between the unique odor of bladder cancer and other bladder illnesses. In the experiment, the dogs had to differentiate between seven types of urine that were placed in petri dishes. Subsequently, they had to lie down in front of the dish from the patient with bladder cancer. The dogs chose correctly in 41 percent of the tests, which is far more than could be expected had they simply been choosing at random (14 percent).[42] It is of interest to note that during the training period, one of the dogs repeatedly chose the urine sample from one of the presumably healthy participants. A more thorough test showed that this person had an undetected—until then—and life-threatening tumor in one kidney.

In 2004, fifteen years after the dog of the forty-five-year-old woman from London discovered the cancerous mole and chemical substances indicating cancer were found in the woman's blood and urine,[43] the

American dog trainer Duane Pickel and colleagues in Tallahassee, Florida, speculated about whether dogs could detect these substances.[44] The first trials used two dogs, a schnauzer and a golden retriever. One of the dogs confirmed cancerous moles in six out of seven patients. The other dog investigated four of these seven patients and reacted like the first dog. In another experiment—carried out in 2006 by the cancer research scientist Michael McCulloch at the Pine Street Foundation in San Anselmo, California, and his colleagues—five dogs (Labrador retrievers and Portuguese water dogs) were presented with sealed test tubes containing the breath of humans with and without lung cancer and breast cancer. The dogs' sniffing detection results were extremely accurate (99 percent for lung cancer and 88 percent for breast cancer).[45]

On a global basis, more than 205,000 women are diagnosed with ovarian cancer annually. Only 46 percent of these are still alive five years after receiving the diagnosis.[46] Chief physician György Horvath performed several cancer experiments with his dog at the Sahlgrenska University Hospital in Göteborg in Sweden in 2008, and his dog identified 100 percent of the tissue samples from patients with ovarian cancer and 98 percent of the tissue samples from patients without cancer correctly.[47] Horvath has several dogs, and two years later two more of these succeeded in detecting these substances in blood samples (plasma) from patients with ovarian cancer. The dogs scores were 100 percent correct.[48] Horvath's dogs did not recognize other cancer odors, which would imply that all types of cancer have their own specific odor. There is also a specific odor for colorectal cancer that is produced when cancerous substances are circulating through the body. A Labrador retriever was allowed to sniff breath samples and feces samples from patients with colorectal cancer and from healthy individuals. On the breath samples, the dog scored 91 percent, for the feces samples, 97 percent.[49]

The urologist Jean-Nicolas Cornu at the University of Paris and his colleagues used dogs to detect prostate cancer in 2011—after having trained dogs from the Belgian sheepdog breed (Malinois) for two years. The dogs were trained to recognize whether urine samples came from men with prostate cancer or from healthy men. The research scientists investigated the urine samples from 66 people for the trial—33 with prostate cancer and 33 without. The dogs succeeded in 30 out of 33 cases (91 percent) in identifying the urine samples from the cancer patients. What remains to be done in this context is identification of the chemical compounds that create this characteristic cancer odor.[50]

The majority of the experiments show that the use of dogs is a promising method for early cancer detection, but further challenges remain.[51] Both the methods for training dogs and for storing the samples need to be improved.[52] Moreover, the research scientists Emily Moser at New College in Sarasota, Florida, and Pine Street Foundation's McCulloch hold that the control samples and cancer samples should come from individuals who are about the same age,[53] but not all research scientists agree on this, since dogs will not be influenced by individual personal odors if they have been trained properly.[54] The odor of the hospital should be avoided, since the dogs can become accustomed to this odor, which can lead to more false outcomes. Breath samples should therefore be collected from outside the hospital.[55] Many people perceive the scent of phenol, for example, as being a typical hospital smell, since phenol is employed as a disinfectant.[56] It is important to be aware that both the ability to detect odors and the length of concentration span do not just vary from dog to dog, but also for the same dog from one day to the next.

Different breeds also have different genes that influence their sense of smell for certain substances.[57] Skeptics hold that using dogs to detect cancer can become an overly complicated procedure, compared with chemical methods in the laboratory, such as mass spectroscopy.[58] Others maintain that electronic noses (an apparatus containing sensors and a detection system that senses odor molecules) are better.[59] When it comes to bladder cancer, the electronic noses are the best. In other cases, the methods are equally effective, while the dog's nose is better than the electronic nose for ovarian cancer and breast cancer.[60] Doctors have pointed out that in cases where ovarian cancer is detected at a late stage and the prognosis is bad, the development of cancer takes place so rapidly that having a dog with a keen sense of smell does not help.[61] In a study from 2013, Horvath and colleagues demonstrated that the dogs were able to detect ovarian cancer in patients at an early stage. The cancer could not be diagnosed using other methods before two to three years down the road. Use of a dog can thereby without any doubt increase cancer patients' chances of survival.[62]

There is reason to believe that the use of dogs holds a large potential, for example, in the context of searching through large crowds in poor nations without the resources to carry out advanced laboratory tests.[63] Moreover, African American women are much less likely to go to the doctor for breast cancer exams than women from other population groups, which makes early detection of a cancerous tumor in these cases unlikely.[64] Using dogs as "oncologists" is an interesting method and is also quick and painless.[65] Breath and urine samples can be taken anywhere and sent into the laboratory.[66] Breath samples have a large potential for the detection of different types of gastrointestinal cancer, such as stomach cancer, intestinal cancer, rectal cancer, liver cancer, pancreatic cancer, and gallbladder cancer.[67] Patients would not need to travel to special clinics since no X-rays or manual palpations of the breast in an exam are involved. There is little doubt that these results are encouraging, and future research will hopefully show whether the dog's sense of smell can be an effective tool in terms of early cancer detection.

In 2006 to 2009 in Norway, Turid Buvik at the Trondheim dog training school, in collaboration with research scientists from St. Olav's Hospital, trained dogs to detect lung cancer in ninety-three patients using their sense of smell. They tested both breath and urine samples, and the dogs distinguished between samples from cancer patients and healthy patients in 99 percent of the pilot experiments. However, the dogs were not as successful when tested in distinguishing between people with malignant and benign tumors. In these tests, the dogs had an accuracy of 56 to 76 percent. The specificity—how often a dog does not indicate a can that does not contain the odor of a malignant tumor—was also very low (8.3–33.3 percent). The research scientists therefore concluded that the dogs were not good enough to be used in the medical examinations of lung cancer patients in clinics.[68]

Since the summer of 2011, Siri and Jens Stedje, Hogne Hole, and Henriette Schermacher Marstein have trained dogs to give indications for the scent of prostate cancer in the urine of patients. Many men develop this form of cancer every year. The project's objective is to develop a precise, cost-efficient, and painless method for early detection of prostate cancer. If the cancer is detected at an early stage, the patient's potential for survival can be increased by 30 to 40 percent. By working in collaboration with private stakeholders, instructors, and dog handlers from the police force and search and rescue dogs, along with the Prostate Cancer Federation, the hope is that as many people as possible will take part in developing this method. The dogs from the project have responded to prostate cancer tumors that were only 1 millimeter in size. They also train dogs to detect other kinds of cancer, so people who have recently been diagnosed with lung cancer, breast cancer, prostate cancer, or cervical cancer and have not yet received treatment can take part in the project. For example, a man around sixty years of age was hospitalized because his urine contained more blood than urine. He was discharged the following day after having been diagnosed with a renal pelvic infection. His wife was not convinced by the diagnosis and contacted with a group called Cancer Detection Dogs. The dogs gave indications for cancer, and Sissel Overn, the urologist who collaborates with the group, confirmed that the man actually had bladder cancer. A short time after the dogs had given indications for cancer, the man received treatment. He is eternally grateful to the dogs—their detection led to diagnosis of cancer at an early stage, which in turn resulted in rapid intervention and treatment. He was given a clean bill of health and is back at work full-time. Another man, whose doctor had informed him that he had all the symptoms of prostate cancer, also had Cancer Detection Dogs check a urine sample. It turned out that the dogs did not react to it. The man followed up by having his doctor do further tests, including a biopsy. Months of uncertainty subsequently followed: Did he have cancer, as his doctor had believed, or were the dogs right? The answer was that he did not have prostate cancer.[69] The dogs were right!

Dogs Can Detect Other Illnesses and Allergies

There are many cases of people reporting that their dog reacts to epileptic seizures, pregnancy, or if they are about to faint or have a migraine episode.[70] There have also been claims that dogs can give warnings about heart disease and when someone is on the verge of falling into a coma. It has been proposed that dogs are reacting to odor,[71] but it could also be the person's behavior that causes the dog to react. There are still no scientific studies that support the hypothesis that it is odor that causes dogs to react and warn their owners.

"Peanut dogs" react to different products that contain peanuts and other tree nuts and are enormously helpful for many people who have nut allergies.[72] Meghan Weingarth of Suwanee, Georgia, is allergic to nuts, and her life has changed because of her dog LilyBelle, a goldendoodle (golden retriever/poodle mix), accompanies her every day.[73] LilyBelle was trained by Ashleigh Kinsley from the Georgia K9 Academy to give indications for foods containing nuts.

Seven-year-old Kaelyn Krawcyk from Durham, North Carolina, has a rare cell disease that can cause a serious and rapid allergic reaction that can be fatal. Her dog JJ reacts to this serious allergy, so when Kaelyn needed to have surgery at Duke Medical Center, JJ was allowed to accompany her into the operating room. JJ is able to detect an allergic reaction five minutes before any vital signs are visible, something available medical technology was not able to do.[74] Dogs are also used to detect bacterial infection caused by the bacteria *Clostridium difficile*, which causes severe diarrhea. This infection is particularly dangerous for patients with a debilitated immune system.[75]

We have only just begun looking at the beneficial opportunities to be found in using the dog's fantastic nose when it comes to detecting illnesses. Tuberculosis is an illness caused by bacteria, and several chemical compounds caused by the illness have been identified.[76] In 2011 a number of research scientists discovered that honey bees and rats can smell this illness.[77] Claire Guest, research scientist and head of Medical Detection Dogs in England, has proposed training dogs to detect badgers with bovine tuberculosis, so individual infected animals can be eliminated.[78] Up until 2014, unfortunately, the lives of many healthy badgers were sacrificed because of the disease. Other conceivable tasks can be to train and investigate whether dogs can sniff their way to the detection of different kinds of venereal diseases, the Ebola virus, bird flu, and tick-borne diseases.[79]

Dog Training Organizations

Dogs are now systematically trained by organizations that specialize in illness detection. Among these are Support Dogs and Medical Detection Dogs, both in England; and in the United States, Dogs 4 Diabetics and the Wildrose Diabetes Alert Dog Foundation.[80] In 2009 the abovementioned Claire Guest received confirmation that she had breast cancer when she trained her first cancer detection dog. Daisy jumped up on Claire and pushed her nose into her breast, which was abnormally sore, and upon closer inspection, she discovered a lump.[81] Medical Detection Dogs began working in 2010 with prostate cancer detection from urine samples,[82] the year before they started working with patients with narcolepsy (sleepiness), cataplexy (sudden, transient episodes of muscle weakness), asthma, nut allergies (walnuts, cashew nuts, and peanuts), and Addisonian crisis (acute adrenal failure).[83]

11

Field Assistant

Research scientist Donna Shaver is working for the National Park Service in Texas to save the world's rarest sea turtle.[1] The Kemp's ridley sea turtle was at one time a species commonly found in the Gulf of Mexico, but due to shrimp trawlers, fishing nets, oil spills, and vehicles on the beach, biologists feared that it would become extinct. In 1985 there were only 702 nests left in the entire world, and the majority of these were on a 25-kilometer-long beach in Mexico. In the past twenty-nine years, Shaver, her colleagues, and volunteers have worked to establish a stable sea turtle population on Padre Island in Texas. The nests of this sea turtle are much more difficult to find than those of other sea turtles, since it prefers to lay its eggs during the worst of storms. A storm will erase all traces of the mother when she comes up out of the ocean to bury her eggs on the beach. In the 1980s, biologists discovered a nest every other year to every third year; they found five nests in 1995 and fifty in 2005. In the same year, a cairn terrier named Ridley was born. Shaver decided to teach him to find the nests containing the newly laid eggs, and in 2012 they found a total of 209 nests. After Ridley has found the nests, they are dug up and the eggs taken to the laboratory, where they are hatched under controlled conditions. The small sea turtle babies are later released into the Gulf of Mexico. Several hundreds of sea turtles have survived thanks to Ridley's outstanding nose work. If Ridley hadn't found all these nests, predators, cars on the beach, and other human activities could have substantially reduced the number of sea turtle eggs that would have hatched and the number of babies that survived.[2]

Dogs Assist Ecologists, Conservation Biologists, and Wildlife Managers

Dogs have been given increasingly more work tasks in research, such as helping ecologists, conservation biologists, and wildlife managers with different field tasks.[3] They work in different types of environments, such as the ocean, deserts, steppe zones, jungles, and forests.[4] Dogs are used to find live animals for tagging in research projects, in eradication programs using trapping and hunting that target unwanted harmful species, and to locate "native species" that are rare or in danger of extinction.[5] They are also used to find animals that have been sedated, dead animals (cadavers), and different types of animal traces, such as feces.[6] These dogs are not commonly owned as pets. They are trained to search for something they normally have no innate interest in and are most versatile if they have no particular interest in prey. They can work on or off a leash, but are supervised by a dog handler at all times. The instruction and training they have received, the target they are searching for, their personality, and the terrain in which they are working all play a part in determining how far away from their handler the dogs will work.[7]

Acquiring information about a species population is extremely important for the proper management of wildlife conservation. Obtaining data about rare and endangered species that are vulnerable to disturbance by humans can be a challenge. In small populations, the traditional methods for capturing live animals are not very effective. The methods are also very time-consuming.[8] However, dogs can be extremely effective in discovering "cryptic species" and species with small populations or traces of these, all of which are difficult for human beings to find. Dogs are involved to an increasing extent in conservation studies and biological monitoring programs.[9] The dogs are four to twelve times better (depending upon vegetation density—the denser the vegetation, the better the dogs are in comparison with humans) than human beings at finding "cryptic species."[10] The dogs also find odorants quickly and often from a great distance, which is a great advantage when out in the field.

Searches for Birds and Mammals

The first documented use of dogs in conjunction with species conservation work took place as far back as in the 1890s in New Zealand. The Irish conservation biologist Richard Henry was the first to do searches using dogs for the last, still-surviving population of the rare bird the kakapo. Subsequently, kiwis were also located using radio transmitters, or they were found in their burrows by using dogs.[11] These bird species cannot fly, and it was important to find them so they could be transported to islands that were not populated by predators.[12] On Stewart Island, dogs found 97 percent of the surviving kakapos during the years 1980–89. The last living kakapo on the northern main island was named after Richard Henry II. It was estimated that this kakapo could have been at least eighty years old when it died. This bird was moved to different islands without predators and where there were other kakapos. They were thereby able to pass on their genes. The kakapo would in all likelihood have been extinct today had it not been for the work done by the search dogs.[13] New Zealand had no native mammal predators, but people have brought in different species of small predators, such as weasels, and rodents, such as rats and mice. Many bird species in New Zealand cannot fly, so the numbers of many of these have been greatly depleted or populations wiped out by predators and rodents.[14] Terriers are most commonly used to hunt these undesired species. Conservation Dogs New Zealand and Ecoworks New Zealand now train dogs to find different protected bird species. The kiwi is a rare bird group in danger of extinction, and dogs are used to find the birds or their nests in burrows. The birds are captured for tagging so they can be monitored, or they are moved to other, safer areas that are free of predators. They are also used to find cats, mustelids, rabbits, and hedgehogs.[15] In a number of countries, research scientists have used dogs to capture live ptarmigan for use in a variety of research tasks.[16]

Many predators use their sense of smell to find the nests of ground-nesting birds, but the odorants of many wading birds and ducks change when they start brooding.[17] Usually they emit volatile substances that can spread a great distance, but when they are brooding, the substances become heavier (in other words, less volatile), which means that they do not spread as far. This makes it much more difficult for predators (and very likely, also dogs) to find the bird nests.[18]

Norwegian Lundehunds Prevent Airplane Crashes

The Norwegian Lundehund Club was contacted in June 2013 by the company Avinor, asking for help in finding seagull eggs at the Tromsø Airport in Langnes. Adult birds can collide with airplanes, in what are called bird strikes, which is the second greatest threat to air safety after terrorist attacks. Normally, the seagulls must be shot and the eggs collected by crews from Avinor; in 2012, 212 seagulls were shot (Avinor has a special license for this). In the summer of 2013, they tested three Lundehunds to determine whether they could be used to find seagull nests and eggs. The dog trainer Merete Evenseth led the dogs. The Lundehunds knew exactly what to do when they were released, which is perhaps not so surprising since the Lundehund had long been used for puffin hunting.[19] The Lundehund has very good control over its bite. They have been bred to retrieve live birds from their nests, and it was important to prevent damage to the puffin in connection with capture, so as to preserve the down. The dogs' task while hunting was to retrieve large young birds and adult birds from nests in difficult-to-access locations and carry them back to their owner. As of 2014, all forms of puffin hunting have been prohibited in Norway.

When Merete reached the search area, the dogs had their noses to the ground and were sniffing their way to the nests containing the eggs. They found the nests and eggs of the common gull, oystercatchers, eider ducks, and terns. The Lundehund Emil primarily retrieved large eggs, while Merete's dogs, Frøya and Gurine, became experts at retrieving small tern eggs. It is very likely that these would not have been found had the crew from Avinor looked for them, since they are so small and blend in with the sand. The dogs carried the eggs one by one to Merete. One of the dogs did this completely of its own volition, while the other two who had not done this before at first wanted to keep the eggs for themselves. Merete was therefore obliged to haggle with them a bit. Not many eggs were broken, since the dogs were extremely careful. On one occasion when two people from Avinor collected eggs, they found 12 eggs in two hours; the Lundehunds found 69 eggs in one hour. In the course of five afternoons, the Lundehunds collected 548 eggs. This represented 548 birds that could have potentially landed on the airstrip and caused accidents. From May 16 to June 12, 2014, they found 386 eggs, and three more young Lundehunds have taken part in the searches. The eggs were delivered to the Tromsø University Museum and will be used in research.[20]

Searches for Snakes

The research scientist Lawrence Kaluber used a dog to find rattlesnakes in the state of Florida way back in 1956.[21] The dog found five hundred rattlesnakes in two years. In 2010 dogs were trained by Dirk J. Stevenson from the Indigo Snake Initiative in the city of Lumber, Georgia, to find specimens of the northern pine snake and the indigo snake in North America. The latter species is not venomous but is one of the largest snakes found in the southern states of Georgia and Florida. It has lost its habitat due to a decrease in the number of gopher tortoises; the snakes actually make their home in tortoise burrows. Dogs can track down the burrows where the snakes are living and where they cannot be seen by humans. Labrador retrievers were trained to find living snakes and cast-off snake skins, so Stevenson could more effectively determine where a snake was living. The dogs were also trained to ignore other snake species, especially the venomous eastern diamondback rattlesnake. In one experiment, the equipages found 100 percent of the skins and 81 percent of the live snakes. They did everything right in 88 percent of the trials that took place aboveground and in 75 percent of the underground trials. The research scientists established whether the dogs were right or wrong by using cameras to check the burrows.[22] The dogs did not do as well when the temperature rose above 23°C. At such high temperatures, they panted more and this affected their sense of smell and consequently their ability to find snakes.[23] In the United States and Australia, where many of the world's venomous snakes live, there are courses available in how to train a dog to detect the odor of different venomous snakes. The dog can thereby serve as a kind of early warning system and protect its owner from dangerous snakes.[24] In the summer of 2014, many dogs were bit by adders in Norway, and a number of places ran out of the antivenin. Dog owners in Norway should definitely train their dogs to recognize the odor of adders.

The Burmese python has become a big problem in the Everglades National Park in Florida. The species is not native to this region, and the origin of the problem very likely stems from irresponsible python owners who released such snakes into the wild in 1979. The population in 2014 was tens of thousands of individual snakes. They eat birds, dogs, and cats, and can also harm humans. The mammal population has been greatly reduced since the snake arrived in the area.[25] Since 2010 research scientist Christina M. Romagosa and her colleagues at Auburn University in Alabama have been using dogs to track down these snakes in the Everglades. The dogs Jake and Ivy, both of whom are Labrador retrievers, have so far helped these scientists capture close to two hundred Burmese pythons.[26]

The brown tree snake was introduced to the American island of Guam in the western part of the Pacific Ocean by accident. When war matériel was transported there from New Guinea after the Second World War, the snake was on board.[27] The snake originally came from the east coast of Australia, East Indonesia, New Guinea, and the Solomon Islands. It is believed that the population on Guam arose from a single female snake.[28] In the course of the 1970s, the native population of birds was wiped out in almost all parts of the island. The snake reproduced rapidly on Guam since it had no natural predators there. Out of the twelve native woodland bird species, only three species survived in the wild. The brown tree snake has also wiped out several species of lizards and has had an impact on the bat population on the island.[29] The snake also attacks chickens, battery hens, newborn piglets, kittens, and puppies.[30] So far the snake has not killed any humans, but it has afflicted children with life-threatening bites.[31] The snake also represents large financial costs since it climbs up power lines and into transformers and causes short circuits.[32] Fewer tourists also come to Guam because of the snake.[33] In the mid-1980s, there were approximately 13,000 to 26,000 snakes per 1.6 km^2, and it was important to locate and contain the dispersal of this alien species as quickly as possible.[34]

Since 1993 Jack Russell terriers have inspected high-risk cargo from Guam to prevent further dispersal of these snakes.[35] This breed of dogs was chosen because it is energetic and tough by nature. Since this dog is so small, it can easily slip into cargo holds and other constricted spaces to do its work. When the dogs arrive at the island, they are trained for approximately four months. When they are fully trained, they are used to inspect outgoing containers, airplanes, and ships.

In a two-year study, biologist Richard M. Engeman from the National Wildlife Research Center in Fort Collins, Colorado, documented that in the first and second years, the dogs found 61 percent and 64 percent of the brown tree snakes put out in containers, respectively.[36] When the dogs checked airplanes that were going to take off from the island, their job was to find the snakes and kill them. For ethical reasons, this practice has now been abandoned and the dogs no longer kill the snakes.[37] It is important to prevent the dispersal of the snakes to other islands where they can wipe out even more birds and other small animals. Up to now, Jack Russell terriers have proven to be the dog best suited for finding snakes in the sites where containers are loaded and before the snakes are able to enter the aircraft. Hawaii is a vulnerable island since there is a lot of air traffic there from Guam. In Hawaii, beagles are used to check cargo and airplanes that arrive from Guam.

It is also important to find brown tree snakes living in the natural environment. Biology professor Julie A. Savidge at Colorado State University led a study in 2011 in which they fed snakes mice containing radio transmitters, so they could later find the snakes without leaving behind the odor of humans on them. Someone later tracked down the snakes and found where they were living, but the equipages were not informed of the snakes' location. The equipages were then given the task of searching a $40m^2$ woodland area and were supposed to find a snake in an area smaller than $5m^2$. In the course of 85 trials, the equipages had an average success rate of 35 percent: they found 30 of the 85 snakes. These experiments demonstrated that dogs can be relatively effective when it comes to finding snakes in daylight,[38] and they are much better than human beings, who only found 7 percent of the snakes at night, even though it is easier to find them at that time of day since they are nocturnal.[39]

Searches for Dead Animals

Research scientists are often interested in locating animal cadavers in order to clarify the cause of the animal's death. This can be a matter of everything from complete animal cadavers to small body parts. The animal may have died of disease, been killed by a predator, or flown into a power line or wind turbine. Searches are often carried out over large areas, which is extremely time-consuming. It is a challenge finding the cadaver before it is removed by scavengers or predators, or before it rots. The best practice is to begin the search as soon as the animal has died.[40]

Searches for Dead Bear Cubs

PhD candidate Sam Steyaert and colleagues at the Scandinavian Brown Bear Research Project used a dog to find bear cub cadavers. Many of the bears in the Swedish forests have been tagged with GPS transmitters. This makes it easier for research scientists to find places where bears have been behaving abnormally or where a female with offspring and a male have met or been in the same place for a long period of time. The dog was used to search for bear cub cadavers around these locations. Adult male bears kill cubs that are not their own offspring, and then mate with the mother of these cubs a few days later. In this way, they pass on their own genes. Usually cubs stay with their mother for one to two years. The female bears defend their offspring, and research scientists often find clear evidence of fights, such as trees that have been broken or scratched, and traces of blood and scraps of fur scattered around on the ground. In some cases, the female bear manages to escape with her young. In other cases, some or all of the offspring of the litter are killed. The female bears can also be killed by males. In the course of a three-year period, research scientists found thirteen bear cub cadavers, and almost half of the eighteen female bears being monitored by the scientists experienced their young being killed or attempts to kill their young. Research scientists also found that female bears with young will use other parts of a given area than those used by male bears and single female bears during the critical mating season. These are areas where the food is of a poorer quality, making it disadvantageous.[41] In Canada, female bears with young also avoid trees marked by male bears during mating season to communicate their dominance to other male bears. When the mating season is over and throughout the course of the autumn, female bears will avoid these trees less.[42] We don't know how the male bear knows that he is not the father of the offspring, but it is very probable that he uses his sense of smell.

Searches for Dead Grazers

The losses of sheep and lambs on rough grazing lands are considerable, and it is extremely difficult to find these cadavers.[43] It is important to determine where and when the animals have been taken by predators. In recent years, dogs have been used in connection with searches for grazers lost to predators.[44]

Cadaver searches using dogs are far more effective than searching for cadavers without dogs.[45] The Great Pyrenees mountain dog has a number of characteristics that make it highly suitable for this purpose. Other breeds can also be used, on the condition that the dog can be safely released into areas with sheep populations.

Searches for Dead Birds

It is not always easy to find bird cadavers when searches are carried out in dense vegetation. In a search done in 2001, wildlife biologist H. Jeffrey Homan at the National Wildlife Research Center in North Dakota and his colleagues found 45 percent of common sparrow cadavers, while the dogs, who had not received any special training before the search, found 92 percent.[46] In 2001 dogs were also used in Norway to search for dead birds and feathers along power lines.[47] In 2006 research scientist Ole Reitan at the Norwegian Institute for Nature Research (NINA) trained and used the feather search dog Luna, a giant schnauzer, to find feathers and other remains from birds that had flown into wind turbines in Smøla. In particular, the remains from eagles—both the white tailed-eagle and the golden eagle—were found, along with remains from willow ptarmigan and wading birds. Luna also found remains from species such as the common raven, the great black-backed gull, the redwing, the hooded crow, the northern wheatear, the greylag goose, the European golden plover, the common snipe, and the common redpoll. Feather search dogs are also used for searches on the Hitra Wind Farm in Norway. By using dogs to locate dead birds, we acquire a lot of information about large and small species, and the distance away from the wind turbines that the remains of dead birds are to be found.[48]

Searches for Bats

Twister, a springer spaniel, finds bat cadavers in and around wind farms.[49] It is very likely that many bats die due to the air pressure surrounding the wind turbines, and the dogs help to establish the impact of the wind farms on the bat population. The dogs are much more adept (73 percent accuracy) than people (20 percent) at finding the dead bats.[50]

Searches for Fecal Matter

The use of dogs for the purpose of finding fecal matter from different species in North America began in the 1970s. These finds gave scientists a substantial amount of new and important information. Fecal matter is the wildlife product that is easiest to find, and there is a lot of it out in the wild. Beyond searches for scat from mammals, dogs can also be used to find regurgitated droppings from birds.[51] Dogs have served an important purpose due to the many opportunities for feces analysis that can be subsequently carried out in the laboratory. By analyzing feces, it is possible to extract an incredible amount of interesting information about the animals:[52]

- the species in question
- the identity and size of the population
- sex and, therefore, sex distribution in the population
- whether they reproduce
- age
- kinship between the animals in the population
- health condition and any diseases
- whether stress is a factor

- what they eat
- physical condition
- the status of the immune system
- any exposure to toxins

If the location of the feces is also recorded, it is possible to acquire information about choice of habitat and the size of the territory inhabited by the animals.[53]

It is expensive identifying species using DNA analyses. The use of dogs can be an option that is less costly than carrying out different analyses in the laboratory.[54] When the dogs find feces, they have already done much of the preliminary work required for various DNA techniques. It can be the only means of determining whether a rare or endangered species is found in an area. By collecting feces, we can therefore acquire much of the same information as from blood tests or hair samples from animals that must be captured, while the degree of risk is much lower for both the animals and scientists.

Bait is usually used for live capture with cage traps, and this often attracts certain animals in a population, such as dominant males. It can also be difficult to find enough samples, and it will be time-consuming in cases where, for example, the terrain is rugged and the vegetation dense. With dogs, this task becomes much easier.[55] It is important to be aware that feces could have been in a given habitat for days, weeks, even months, and that this is not necessarily proof that the animals are still to be found in the area.[56]

Searches for Whales

For some dogs, their place of work is the ocean. They are often put in front at the bow of a small vessel, where they can sniff out feces from North Atlantic right whales.[57] It is important to find these feces quickly, because they will sink in the course of half an hour. In 2012 conservation physiologist Katherine L. Ayres and colleagues at the Center for Conservation Biology at the University of Washington (UW) used the black Labrador Tucker to collect feces from orcas (also known as killer whales), with the objective of subsequently establishing in the laboratory the levels of stress and hunger hormones.[58] Orcas are difficult to study because they are underwater almost 90 percent of the time. Tucker makes the job much easier: he can detect feces from an unbelievable distance of 1.8 kilometers. The feces have a fish odor that is unique for endangered southern resident killer whales found in the south in British Columbia, Canada, and the state of Washington. When Tucker is situated at the front of the vessel, he sniffs out floating whale scat. He does not jump into the water when he detects the feces, but gives a clear indication of a find, so the research scientists can fish it out using a large landing net. The feces are put on ice and taken to the laboratory for hormone analyses. In the course of six months, Tucker found 150 feces specimens. The research scientists were subsequently able to establish a strong correlation between the salmon shortage and the shrinking population of orcas. It was first believed that the decrease was due to the many whaling safari boats also to be found in the area. The southern resident killer whale population decreased by 20 percent from 1995 to 2001. Toxins are stored in the body fat, and when food supply is poor, the orca consumes the stored body fat. The result of this is that toxins are released in the body. It is therefore important for the whale to have adequate access to salmon in the springtime to ensure their chance of survival.[59] Tucker has also been trained to find droppings from wolverines, wolves, moose, and reindeer.

Searches for Predators on Land

Predators that live on land are sensitive to disturbances and the splitting up of their territories by human activity and are often difficult to study. They move across large land areas, have a naturally low density, and, behaviorally speaking, are shy. In 2007 conservation biologist Robert A. Long at the University of Vermont and his colleagues wanted to monitor different predatory animals in a given area. They used dogs to search for feces from the black bear, fisher cat, and bobcat and could thereby collect DNA for analyses. The research scientists also wanted to compare how wildlife detection dogs measured up against two other methods that are often used in the study of predators, specifically camera trapping and hair traps. They found that when using dogs, the probability of finding a black bear was 87 percent, 84 percent for the fisher cat, and 27 percent for the bobcat in one study,[60] and, respectively, 86 percent, 95 percent, and 40 percent in another study.[61] Because of the dogs' effectiveness, this monitoring method proved to be better and less expensive than camera trapping and hair traps.[62]

The Siberian tiger is another species that is extremely difficult to monitor. Genetic analyses have proven to be a bad solution due to the tiger's low genetic variation. The traditional method for monitoring Siberian tiger populations in Russia is to look for trails in the snow. In Russia in 2007, dogs were trained by biologists Linda L. Kerley and Galina P. Salkina at the Lazovsky Nature Reserve to identify individual Siberian tigers from feces picked up while tracking in snow.[63] The feces were taken to the laboratory, and in one trial the dogs had to find the right specimen in a circular layout. First, the dogs sniffed at the feces from the tiger, and then they had to find it again in a layout of specimens that also included feces from six other tigers. The dogs correctly identified the feces in 87 to 100 percent of the attempts, depending on the type of test being done. The research scientists could thereby also map out the movements of the different tigers and determine the size of the areas they inhabited. It is very important to know where different individual animals move about in areas when human beings constitute a threat for this critically endangered species. Such mapping can therefore provide important information about how to manage the population. This method can also be a good alternative to genetic methods in feces studies for which DNA analyses are impractical or not very effective.[64]

It can be difficult to distinguish visually between the feces of different species because many animal species produce very similar feces. What is believed to be feces from a bear turns out to be from a badger, while wolf feces can be easily confused with dog feces and those from a fox confused with a marten's. There are dogs that have been trained to distinguish between the feces of different species of fox—such as the kit fox, red fox, and gray fox—or between the grizzly bear and black bear.[65] Even bear scientists cannot do this with certainty without performing expensive laboratory analyses. The dogs managed to find all the feces samples from the rare kit fox when red fox feces were also present, but they were not as good at ignoring red fox feces in the absence of kit fox feces.[66] Genetic tests were also carried out, and these disclosed that all 329 feces samples the dogs had found were from the correct species of fox. The dogs also found four times more kit fox feces than an experienced person would have managed. In 2005 the dog of wildlife biologist Deborah A. Smith at UW's Center for Conservation Biology located 825 kit fox feces samples.[67] All of these were from this species (confirmed through genetic experiments), even though there were also feces from other species such as the coyote, skunk, and American badger in this area. If the monitoring takes place in the course of the reproduction period, it is important to remember that the animals can change their marking pattern. When the pups of the kit fox are born, the parents change the pattern in which they deposit feces. This makes it easier to find male feces and more difficult to find feces from females, and on the whole can lead to underestimating the number of animals in a population.[68]

Searching for Badger Feces

In Europe there has only been one scientific study in which dogs have been used to find fecal matter from birds and mammals, and that was in 2012, in Finland, where dogs were trained to find the feces of raccoon dogs and badgers.[69] This stands in strong contrast to the situation in North America, where a series of successful studies have been carried out. Telemark University College is now collaborating with the University of Oxford on a large-scale badger project in Wytham Woods in Oxford, investigating how badgers communicate using odor signals. One of TUC's collaborating partners, Chris Newman, used dogs in the context of the work on his doctorate to find badger latrines (defecation sites). The year he did not use his dog, Samson, he found 178 latrines; the following year, he found 348 with Samson's help.

Challenges

It is important that sniffer dogs are able to focus on finding feces solely from the species of interest. The dogs will learn to search for feces from other species if the dog handler shows an interest in feces that appear similar to those of the species of interest. If the dog's companion shows an interest in feces from other, similar species, this can cause the dog to sit down in expectation of a reward. The handler might then believe that the feces are from one of the species being searched for, which will increase the probability of the error being repeated. Training with few feces samples or low-quality feces can also reduce accuracy. In 2004 research scientist Carly Vynne at UW's Center for Conservation Biology and her colleagues had only six to eight training samples available from the species they were studying. These were stored together with samples from other species, and for that reason there is a possibility that the feces from the species to be searched for were contaminated by these other samples. It is worth noting that dogs will more frequently find feces from common species, and they are therefore rewarded more frequently for these finds. Because this might make them less interested in finding the scent of rare species, the dogs must be trained regularly with feces from less common species.[70] If there are some species for which researchers don't want indications, the dogs should be trained for this as well. Using a training platform, the researchers can, for example, present the wanted and unwanted feces side by side and only reward the dog when it finds the right feces.

Scandinavian Brown Bear Research Project

In 2007 the head of the Scandinavian Brown Bear Research Project,[71] Professor Jon E. Swenson, and postdoctoral research fellow Andreas

Zedrosser were interested in learning more about the odor-based communication of brown bears. I was contacted and invited to take part in the project. The history of the Scandinavian Brown Bear Research Project extends far back in time—it was started in northern Sweden in 1984. An area of study in the south was added in 1985, and since 1987 it has been a joint Scandinavian project. Fieldwork is done in Dalarna, Gävleborg, and Norrbotten Counties in Sweden and in Hedmark County in Norway. The primary objective has been to understand the ecology of the Scandinavian brown bear, to develop a scientific knowledge bank for management of the species in Sweden and in Norway, and to provide information for the general public.

The research scientists knew little about bears' communication with odors in 2007. It had not been established, for example, whether or not the bear had anal sacs. Four years later, we discovered that it does, and using chemical analyses, we also found that the anal secretions of the two sexes are different.[72] In 2012 we discovered that young bears living in a zoological park could distinguish between unfamiliar adult male and female bears' anal secretions.[73] We do not know, however, how this secretion is used in the bear's natural habitat.[74] It would be very exciting to train some dogs to find out when and where the brown bear uses this secretion and whether it is deposited beside their feces. The odorants of feces and anal secretions are different,[75] so it should be easy for a dog to distinguish between these. And what kind of information is found in the anal secretion, in addition to information about the sex of the bear? What about the urine? Where do the bears deposit markings? Are there individual differences in the bears' urine and/or anal secretion markings? Can we use dogs to find out whether a female bear is pregnant? Can the dogs sniff out different illnesses in the feces of bears? These are just a few of the future possibilities. PhD candidate Melanie Clapham at the University of Cumbria in England has investigated the brown bear's (grizzly bear's) scent-marking activity on trees in Canada using camera traps.[76] Adult male bears sit down beside trees and possibly deposit anal secretion at the root of the tree. It would be interesting to train dogs to determine whether they actually do this.

Tracking Down Plants

In the United States, dogs have been trained to search for non-native plants, because such plants can displace native species and pose a threat to biological diversity.[77] Such changes can have large financial costs, so it is important to eliminate these alien plant species as soon as possible. All plants produce a mixture of volatile organic compounds, giving each species a characteristic odor, and it is possible to train dogs to find a plant species or plant family. While it is difficult for human beings to find small, young plants, for dogs this is a much easier task.[78] When dogs have been trained to find a rare species of knapweed (*Centaurea*), they are more accurate than people and are especially adept at finding the small plants.[79] In the United States, there are dogs able to sniff their way to an introduced species of Chinese clover and to a rare lupine flower that is the host species for an endangered butterfly.[80] By finding and potentially protecting this lupine, conservation biologists hope that the butterfly will also be saved. Dogs are also used to find an endangered species of the wood rose in New Zealand and to find plants that are stolen from public spaces and roadsides in the United States.[81] Searches for many different types of mushrooms are done using dogs in many countries (see chapter 13, "Other Work Tasks for Sniffer Dogs"), but there is also a great potential for many other plant groups.

Advantages and Disadvantages of the Use of Dogs as Field Assistants

One disadvantage of using dogs as field assistants is that they need a lot of training, which can be very expensive. The dogs are very agile, but their handlers cannot always access demanding terrain in the same way. Different individuals and breeds can be better suited to find one species than others. Some individual dogs can be very proficient at finding bears, but if they are given cat feces, they would rather roll in them. Dogs can also transmit illnesses into the wild,[82] and it is therefore important that they are checked out by a veterinarian and fully vaccinated, and that one is attentive to any changes in the dog's health. Dogs can also disturb wildlife.[83] In 2007 Peter B. Banks, professor of conservation biology, and research scientist Jessica V. Bryant at the University of New South Wales discovered that the diversity of birdlife and the number of birds decreased by 35 and 41 percent, respectively, in a wooded area north of Sydney, Australia, when people walked their dogs there.[84]

It is important to be aware that the use of dogs to locate animals in the field can increase the risk of attracting predatory animals or changing the behavior of the species being searched for. These potential, negative effects become even more problematic when it is a matter of an endangered species. Led by research scientist Jill S. Heaton at the University of Nevada, Reno, a 2008 study on the Mojave Desert tortoise, however, showed that dogs did not represent an increased risk of the tortoise being eaten by predators. Neither did the dogs' presence change the tortoises' movement patterns more than when human beings did searches alone. No tortoises were taken by predators in the course of the year following the search.[85] In some cases, the presence of dogs, or their odor, can in fact frighten away predatory animals.[86] The dogs' well-being is also important. If the dogs are at high risk of injury or being killed by other animals such as bears, wolves, or venomous snakes while on the job, it is best to find other search methods. In parks and areas frequented by a lot of people, the use of dogs is not ideal, because the dogs might be disturbed in their work.

Dogs Can Be Trained to Search for Many Different Animal Taxa

Insects

In 2011 and 2012, articles were published in England about dogs that had located up to thirty-three hives inhabited by four different species of bees (the large carder beoe, the great yellow bumblebee, the red-tailed bumblebee, and the heath bumblebee).[87] The dogs sniffed out the materials of the different nests.

Mollusks

The rosy wolf snail was brought to Hawaii in the 1930s to keep a large African snail under control, but instead ended up eliminating three-fourths of the native snails of Hawaii.[88] Dogs are now used to find this snail in order to eradicate it.

The California Department of Fish and Game was the first to train a dog to find zebra mussels in 2007. The species was not native to California, and the dog Popeye is now helping to prevent its propagation.[89]

Fish

Dogs can sniff out fish that have been caught illegally and determine whether water samples contain odorants from fish that do not belong in a particular water system.[90]

Amphibians

Dogs can sniff out rare frogs and salamanders.[91] In England, dogs find natterjack toads and large salamanders. The dogs are trained to recognize the feces and tail of an endangered salamander species in order to find both sexes of living salamanders.[92]

Reptiles

The dog Apple finds live reptilian species. In New Zealand, Apple has been used with great success to find rare reptilian species such as tuataras, green tree geckos, and wood geckos.[93] Dogs can detect tortoises from more than 60 meters away.[94]

Birds

In addition to live and dead birds, dogs also find eggs and feathers.[95]

Mammals

In the 1960s, Norwegian elkhounds were used in the United States to find sedated moose and deer that had escaped from research scientists before the tranquilizer had begun to take effect. When the research scientists had dogs with them, they could reduce the dose of the tranquilizer and thereby also the mortality rate. The research scientists estimated that nineteen out of forty-eight animals would have died without the help of the dogs.[96] Today's methods of sedation, however, are much improved.[97]

Dogs have been used to find the feces of a number of species in many different habitats,[98] such as mink and otter,[99] wolf, coyote, wolverine, bobcat,[100] puma, black and brown bear,[101] tiger,[102] leopard, snow leopard, skunk, raccoon, badger,[103] fox,[104] bats,[105] and marten.[106]

Dogs can find scent markings out in the field.[107] Deer fawns lying on the ground, out of the dog's sight, are nonetheless detected by the dog. This indicates that the fawns give off an odor.[108]

Dogs can find places inhabited by a range of different mammals, such as the sleeping sites of bats,[109] black-footed ferrets living in the burrows of prairie dogs,[110] and mammoth remains located under the permafrost in Siberia.[111]

Dogs can find the haul-out sites and breathing holes of the ring seal from a distance of 1.5 kilometers even with wind velocities of up to 46 km/hour (12.8 m/second) and in snow depths of up to two meters.[112]

12

Pest Detector and Building Inspector

Wherever there are human beings, there are rats. In many urban areas, rats represent a huge problem, such as in New York City. They live in the sewer system and in sites filled with garbage and other debris. For this reason, the members of the Ryders Alley Trencher-fed Society (R.A.T.S.) trained their dogs to sniff out rats and kill them.[1] The dogs work in teams. One dog will sniff out the rat and start barking, and another will catch it when it tries to run away. As far back as 1851, terriers were used to catch rats in London. There is even a special American breed of dogs called the rat terrier. In the United States, there is also a sport called Barn Hunt, where dogs have two minutes to sniff through a hay bale labyrinth and subsequently indicate where they smell a rat. The rats are not hurt because they are lying safely inside a pipe. The dog sport even has its own Facebook group that calls itself the Barn Hunt Association.

Pest Detector

Dogs are able to find insect pests much more easily than human beings. Unlike us, they are able to detect the scent of these small creatures. They are much quicker than we are and have an easier time accessing narrow spaces, which can save us a lot of work. The dogs do an important job and can prevent insect pests from causing large problems. Some insect pests suck our blood, while others damage vineyards, fruit trees, or other valuable trees. Some cause internal or external injury to other animals. Using dogs to find insect pests can also have financial benefits because once the insects are eliminated, there is naturally less risk of their returning.

Bedbugs

The hotel is beyond reproach, it is a five-star establishment, but why are there so many itchy red spots all over my body? More and more people find themselves asking this question, and it is very likely due to a case of bedbug bites. Bedbugs are attracted by our body warmth, by the carbon dioxide we exhale, and by some odorants, such as aldehydes and ketones, which are emitted from our skin.[2] In recent years, bedbugs have become a nuisance at many hotels worldwide, and even the cleanest hotel room can be afflicted by this insect pest. In New York City, the number of reported cases of bedbugs increased from 537 in 2004 to almost 13,000 in 2010. The increase is due to the upsurge in travel to places in the world where bedbugs are more common.[3]

Bedbugs are also found in camping cabins, student dormitories, and cinemas, and in our homes, they will hide first and foremost in beds or couches but also other places.[4] Regardless of where they hide, we can wake up with insect bites in the morning. Much of the increase is due to the ban on the use of the pesticide DDT (dichlorodiphenyltrichloroethane) on insect pests, which went into effect in 1970. We travel more, which also increases the risk of bedbugs accompanying us on our journey as stowaways. They can "hitch a ride" on our clothing, bedding, or in our luggage. In a number of countries, insecticides are used to kill bedbugs, and hotel rooms are routinely sprayed with it. In 2011 eight people died at a hotel in Thailand where rooms had been sprayed with the potentially lethal toxin pyrophus, the indoor use of which has been prohibited in many countries.[5]

Books can also spread bedbugs. The insects can "hitch a ride" on books that are brought in and out of libraries. Bedbugs have become a growing problem in a number of libraries in the United States. If you have borrowed a book from the library, you should look closely for tiny insects and droppings, which look like black blotches or ink marks.[6]

Dogs have been trained to find these insect pests in England, the United States, Australia, Thailand, and Norway, because it is difficult for humans to detect them with the naked eye.[7] It is important to discover these pests early and before the population becomes too large. Bedbugs are only 4–5 millimeters long and 3 millimeters wide. A female bedbug will lay 1–5 eggs a day and can produce 200–500 eggs in the course of her lifetime.[8] The eggs hatch after around ten days. Bedbugs have scent glands, which give them a very characteristic odor that a dog can easily detect. In 2008 research scientist Margie Pfiester at the University of Florida and her colleagues trained dogs to detect bedbugs. The dogs found 98 percent of the live bedbugs in the hotel rooms. The dogs were able to find them even if there was only a single insect in the room. The dogs were also able to distinguish bedbugs from other species such as cockroaches, termites, and ants. They could also differentiate between bedbugs and eggs and other bedbug remains, such a shell remnants, droppings, and dead bedbugs. Hotel directors should—before they receive any complaints from guests—have dogs inspect hotel rooms on a regular basis.[9] Dogs will also react to filter paper containing only the alarm pheromones of bedbugs.[10] If the bedbugs are located far above the heads of the dogs and the airstream draws the scent upward, the dogs can miss the insects. It is this kind of error that has introduced some doubts about the effectiveness of using dogs for this task. But experience has shown that in many situations dogs will find insect pests that would have been very difficult for us to detect.

Spruce Bark Beetle

The spruce bark beetle is a large problem for forest workers in many places. The organization SnifferDogs in Sweden has trained dogs to recognize the odorants produced by spruce bark beetles. The beetles attack live spruce trees and also spread fungal spores.[11] In the winter of 2008, dog trainer Annette Johansson from the Swedish University of Agricultural Sciences started training the dog Meja on synthetic odorants from spruce bark beetles. In the summer of 2009, she tested Meja in the field, and Meja succeeded in finding infected spruce trees. Johansson has also trained the dog Aska, and both dogs detect four different types of pheromones from this beetle, which means that damage can be discovered at an early stage. The dogs can sniff their way to the infested spruce trees starting from distances of more than 100 meters away.[12]

Other Insect Pests

The research scientists William E. Wallner and Thomas L. Ellis at Michigan State University started using German shepherds back in 1976 for the detection of gypsy moths,[13] another insect pest. Out in the field, they found eggs on stones, trees, and bark, and the dogs were able to detect the eggs from two meters away. The screw-worm fly is another bothersome parasite. When it matures, the females fly around searching for animals or humans with open cuts or sores on their bodies, laying hundreds of eggs along the edges of such sores. After a few hours, the eggs hatch and the larvae appear. The larvae burrow into the skin, where they have a feast. It is extremely unpleasant to have such larvae in a cut or sore. Both animals and human beings can die from this type of larvae attack. The screw-worm fly can also be the cause of large financial costs. The insect has been wiped out in North America all the way to the northern border of Guatemala and Belize. It is important to find these insect pests to prevent them from returning to regions where they have been eradicated. Inspections of pets, for example, at airports in countries that are free of this insect pest are therefore very important. Sniffer dogs can also be used to inspect humans, vehicles, and animals at quarantine stations. During one trial carried out in 1990, the dog of research scientist John B. Welch from Screwworm Research in San José, Costa Rica, found 95 percent of the animals that were infected by screwworms.[14]

Triatominae, also called kissing bugs, are bloodsucking insects that spread parasites through the blood, which are in turn the cause of Chagas' disease.[15] In Latin America and the United States, 16–18 million people suffer from Chagas' disease. It has been speculated that the British natural scientist Charles Darwin may have died from Chagas' disease, since he was bit by this bug, but more recent findings indicate that he suffered from lactose intolerance.[16] Kissing bugs are nocturnal and often reside in cracks in house walls. They can be dangerous since they often sting near the eyes.[17] The dog of research scientist Miriam Rolon from Centro para el Desarrollo de la Investigación Científica in Asunción, Paraguay, found kissing bugs predominantly in fallen trees. In addition to detecting the adult individual bugs and their nymphs, the dog also found three other bug species. The dog had never been in contact with these other bugs before, which means that both the nymphs and the three other species produced the same odorants as the adult individuals the dog had been trained to recognize. However, the dog did not recognize the droppings of these bugs.[18] In North America, dogs can also detect the Asian long-horned beetle, which causes damage to a variety of hardwood trees and bushes.[19]

The red palm weevil's larvae can cause extensive damage to palm trees in tropical regions in Asia. This weevil spread to Spain in 1994 and was discovered in France in 2006. Six months can pass from the time of the attack until the larvae kill the trees. The larvae live on the soft plant tissue inside the trunk, excavating tunnels inside the palm tree. In the event of a powerful infestation, the trees are weakened and eventually die. Usually the damage is discovered too late. It is extremely important to discover the damages at an early stage so the correct measures can be implemented to curb the damage to palm tree plantations. Two golden retrievers were trained in 2000 by the research scientist Joseph Nakash and his colleagues from the Peres Center for Peace and Innovation in Israel to recognize the scent of a brownish-yellow secretion that is emitted by the palm tree when it has been attacked by these larvae. To test the dogs, the research scientists collected this secretion and put it out on other trees on the plantation. The dogs quickly found the trees with an accuracy rate of 100 percent. They also identified infected trees that had not yet been discovered.[20]

Mealybugs living in vineyards along the coast of California have also been the cause of crop destruction. This is an alien species that arrived in California in 1994.[21] The mealybugs spread from country to country through imported plant materials.[22] It is very difficult for wine growers to visually detect these tiny insect pests, which hide under the bark and in the roots of the vines. They attack the grapes in the vineyards and leave behind honeydew (a sugary excretion), which creates the perfect environment for sooty mold and other diseases.[23] This causes the grape leaves to become black and sticky, and the clusters of grapes turn into a gooey and disgusting lump, thereby rendered useless for making wine.

Honeydew attracts other insects such as ants, and mealybugs can also be carriers of grapevine viruses. Since 2005 Bonnie Bergin, who heads the Assistance Dog Institute in Santa Rosa, California, has trained golden retrievers to detect the odorants that the female mealybugs use to attract the males in April and May. The female cannot fly, and the odorants are therefore an effective way of attracting a mate. The training begins while the dogs are still puppies. The mothers receive a synthetic version of the odorant that is spread on their paws before nursing the puppies, so from an early age the puppies become accustomed to the scent of mealybugs. If these insect pests are detected between January and March when they are under the bark, wine growers can treat individual infected trees. They can remove a grape or two and need not treat the entire plantation. The vineyard owners are thereby spared having to use large quantities of insecticide, which is not only expensive but also a threat to the environment.[24]

Termites are a group of insect pests found in North America that are not easy to detect in buildings. There are no termites in Norway, but there are carpenter ants that have similar behavior. These insect pests eat wood materials and can therefore cause large-scale damage that is expensive to repair. Both species often cause extensive damage before they are detected, and it is estimated that termites do damage to buildings equivalent to US$5 billion annually in the United States alone. The pest control management industry started using dogs to find termites back in the mid-1970s.[25] The initial results were not particularly satisfactory, which was probably due to poor training methods and contaminated training materials.[26] In a study done in 2004 led by the entomologist Shawn E. Brooks at the University of Florida, the dogs found 96 percent of the worker termites. The dogs made mistakes in only 3 percent of the trials when termites were not present. Dogs trained to locate one termite species were also 89–100 percent accurate in finding four other termite species. The dogs were also able to differentiate the termites from other insects such as cockroaches and ants, and to detect building materials damaged by termites.[27]

In South Korea, dogs have been trained to detect termites in temples constructed of wood. In the United States and Japan, dogs have been trained to discover termites in trees, wood products, and telegraph poles. In Abu Dhabi in the United Arab Emirates, dogs find termites and other larvae in date palms and telegraph poles. The damages can thereby be repaired before they become too extensive.[28]

The red imported fire ant was originally found in South Africa but spread to Taiwan early in the twenty-first century. This ant is a well-known insect pest in Taiwan, where it destroys agricultural products, threatens the well-being of both domestic animals and humans, and also represents a threat to local species. After a long period of training, both indoors and outdoors, beagles are now very successful in detecting this insect (> 98 percent). The ants are found in the soil, on farms, in greenhouses, and in seaports. The dogs are also able to distinguish this species from other ant species and can find their nests in the wild.[29]

Mold and Bacteria Detector

Different mold species and bacteria can cause rot in woodwork and other construction materials. Spores, dust, and rot decay can cause allergies and other problems for people living in mold-infested houses.[30] People who live in infected houses can also suffer from fatigue, headaches, skin ailments, respiratory problems, and joint pain.[31] The use of dogs to detect damages caused by microorganisms began in the 1980s, first in Sweden, then later in Denmark, Finland, and the United States.[32] The first fungi detection dog in Norway was a Swedish German shepherd who was borrowed to search for rot in telephone poles in 1991.[33]

In Sweden dogs were used in the 1980s to locate rot decay in telephone and utility poles and fungal infestation in buildings.[34] The dogs learned to recognize ten different fungal species.[35] Mold often grows in places where it is difficult to detect. This can be behind walls, in the floor, or behind different utility fixtures. The dogs are very effective at investigating large buildings such as schools or office buildings. They will quickly gain an overview of where the organisms are found, and a building with two hundred rooms can be searched in the course of eight hours. Dogs are therefore used with increasing frequency to find mold fungi.[36] In Sweden they have also been trained to give indications for old poured screed from an extremely damp environment. The poured screed contained fish protein.[37]

Utility Pole Inspector

Field specialist and dog trainer Cato Sletten has checked many utility poles with his dogs. Five dogs were used for this job, primarily Malinois, flat-coated retrievers, and Labrador retrievers. This is an effective method for checking whether the poles are sound or have been infected by wood-decaying fungi. High season runs from the spring (April), when there is no ground frost, until the ground freezes again in the fall (November). Formerly, the utility pole installers inspected the poles by knocking on them and listening for "infirmities." Healthy and solid poles are critical for safeguarding electric power supply.[38] The dogs spend only five to ten seconds determining whether or not a pole has been infected by rot.[39] Sletten's dog Cleopatra, or Cleo as she is called, digs frantically beside a pole when she finds rot. The rot is usually at ground level (from 45 cm underground to 15 cm aboveground), where moisture, fungi, bacteria, and contaminants can enter the pole through natural cracks in the wood or through pathways created by insects.[40]

Sometimes the dogs will find rot as high up as one meter above the ground. If the rot is higher up, the dogs will try to climb up the poles. If the pole is in water or on a stone surface, it will be more resistant to rot infestation. An equipage can inspect from 100–150 poles in one day, depending on the terrain and accessibility. Approximately 4–6 percent of the poles that are inspected are so rotten that they have to be replaced, which costs around US$3,400 per pole.[41] By replacing the poles in time, before they are blown over, a lot of money can be saved. Many customers will also thereby be spared the experience of losing their power supply due to a destroyed pole. A single equipage will check around 10,000 poles annually.[42]

Bacterial and Rot Damage Detector

Brown root rot found in fruit trees and ornamental trees is a problem in populated areas of southern Taiwan. The leaves on coniferous trees turn brown and suddenly die. The fungus *Phellinus noxius* causes this brown root rot disease.[43] *Phellinus noxius* is a typical wood-decay fungus and infects trees that have rot-infested wounds, broken tree branches, or other similar types of damage. Rot is first and foremost a sign that a tree is old. The older a tree is, the more susceptible it will be to rot infections. The fungus's fruit body (reproductive organ) is usually found a bit farther up the trunk.[44] Sniffer dogs specially trained at the National Pingtung University of Science and Technology and dog trainer Wei-Lien Chi detect this disease before the trees fall over and injure people.[45]

Citrus fruits such as the orange, lemon, clementine, tangerine, and grapefruit are susceptible to contraction of a bacterial disease called citrus greening disease (HLB). This disease is transmitted by psyllids (sometimes called jumping plant lice).[46] The psyllid is the cause of large-scale crop damage in US southern states. The psyllid was introduced in Florida in 2005 and California in 2008. The nymphs of these psyllids transmit a pathogen that prevents the normal development of leaves, and infected fruit remains small and discolored. The juice of the fruit contains many acids and becomes abnormally bitter. The fruit retains its green color when it is ripe, but due to its stunted size and poor quality, the infected fruit has no value. There is no treatment for the disease. The only recourse is to remove the diseased tree at an early point in time to prevent spreading of the pathogen.[47] In the autumn of 2010, this bacterial disease was found on imported poinsettias in Norway.[48] In 2010 research scientist Tim Gottwald at the USDA's Horticultural Research Laboratory in Fort Pierce, Florida, began using dogs to identify infected trees. So far the dogs have found 97–99 percent of the infected trees where they have done searches, and they also detected the fruits infected with the disease.[49]

In Alabama the dogs of research scientist Lori Eckhardt at Auburn University also detected small beetles that are carriers of fungi that attack and destroy the roots of pine trees. The dogs Opie and Charm have been trained to detect these fungi, specifically *Leptographium*, *Grosmannia*, and *Heterobasidion*. The research scientists did not want to dig up the tree roots because this liberates substances that can attract more beetles to the area. The dogs, on the other hand, do not disturb the beetles or spread the fungi.[50]

In 2013 dog trainer Sharon C. de Wet from Australia trained the Labrador retriever Baz to detect a bacterial disease in beehives. This particular disease affects the larvae of the honey bee so they die in their cells. The dogs search for the scent of the dead larvae. The bacterial disease is responsible for huge financial losses and constitutes a growing problem in several places in the world. Baz inspected fifty-one beehives. He identified all the beehives that were infected by this bacteria and 76 percent of the beehives that did not have the bacterial disease.[51]

Roundworm Detector

Dogs can help other animals carrying internal parasites. In 2008 the Australian research scientist Kate M. Richards at La Trobe University in Melbourne, Australia, led a project designed to determine whether dogs could be trained to detect roundworms in the intestinal tract of sheep.[52] This involved three different roundworm species, and there was an interest in determining at which stage the dogs first detected the infection. The dogs were tested with nine paper bags containing sheep feces that were not infected and one bag containing infected feces. These were placed randomly in a circle, and the dogs were then allowed to sniff at the bags. The infected bag contained the feces from a sheep with one of the three roundworms or a mixture of all three. In the course of eighty trials, the dogs had an average success rate of 76–80 percent, depending on the species of roundworm. In the tests using all three roundworm species, they had a success rate of 92 percent. When the infected feces were seven days old, the dogs' success rate was 85 percent. This study demonstrated that dogs can be trained to detect roundworm infections in the feces of sheep. If we determine the chemical composition of these scents, we can produce sensitive electronic noses that can effectively detect parasite infections in all the sheep in a flock.[53]

Scabies Mite Detector

Research scientist Samer Alasaad at the University of Zürich in Switzerland and colleagues trained dogs in 2012 to find chamois located in inaccessible areas of the Italian Alps that had contracted a skin disease caused by a parasitic arthropod (the scabies mite). After a few months, the animals will die, often due to a combination of starvation, infection, and dehydration. The odor of animals with scabies is unique. The dogs found 292 dead animals with the skin disease and 63 infected animals. The sick animals were identified, separated from the flock, and captured by the research scientists. Both dead and infected live animals are potential sources of infection for other chamois and other species in the area, including human beings.[54] Scabies mites can cause intense itching in humans but are not dangerous. Many other mammals are also hosts for the scabies mite. Dogs, cattle, pigs, goats, and red foxes often contract the skin disease, and it is the cause of both suffering in the animals and significant financial costs.[55] Dogs can also be used to find red foxes with scabies in Norway. In a long-term study of the red fox in Norway done in 2008, PhD student Rebecca K. Davidson at the Norwegian Veterinary Institute in Oslo and her colleagues found that the red fox adapted to living with this parasite and some even recovered and their health was restored. The physical condition of animals carrying the parasite, however, was much worse than that of uninfected animals.[56]

Future Opportunities

It is possible that dogs can also be used to find and monitor populations of many insect pests beyond those mentioned here, such as the nests of Africanized honeybees (also known as killer bees), which kill many people every year in South America.[57] And what about lice and fleas? Can dogs become the school nurse's new assistant at schools? The population of ticks has experienced powerful growth recently. Perhaps you should teach your dog to recognize the scent of ticks,[58] so they can inspect you when you come home from a hike?

Many alien species of insect pests continue to threaten the native insects.[59] The buff-tailed bumblebee is an example of such a threat.[60] The same can be said for the harlequin ladybird and Argentine ant, both of which are sibling species of the tramp ant *Hypoponera punctatissima*, also known as Roger's ant.[61] Introduced predators may also be more harmful to prey populations than native predators, and they impose intensive suppression on populations of native species. The American mink has been released into the wild in twenty-eight countries in Europe. In many places the mink is a large problem, where it is an alien species that has escaped from mink farms and is blacklisted.[62] The American mink is a versatile predator. It can pose a large threat to ground-foraging birds, rodents, and other species.[63] It eats the young and eggs of birds, and kills and hoards adult birds.[64]

Gas Leak Detector

Mining operations are dangerous because of the risk of gas leaks, and many people have perished in mines. In the past, mine rescue crews would carry canary birds in a small cage down into the mine after a fire or explosion. When the birds stopped singing, this meant it was time to get out of the mine, since it was a signal that the poisonous gas carbon monoxide was found in the air. This is a colorless and odorless gas. The canary birds functioned as an early warning system. They are in fact fifteen times more sensitive to these gases than human beings[65] and would die before the gas affected the miners. Dogs also have a very low threshold for reacting to gas (such as propane and butane), which in many cases can save lives. In many cities such as Stockholm, the use of gas for cooking and heating is prevalent. The dog Aron was trained in the early 2000s to detect different gases in an urban environment. He was able to locate the source of a gas leak when the smell of gas was reported in large residential areas.[66]

The quality of gas pipes is improving all the time, but leaks still occur. Finding the tiny holes has been difficult, and for that reason in 1991 the research scientist L. R. Quaife and his colleagues at Esso Resources Canada Ltd. in Calgary developed a system that located such leaks.[67] This system consists of two components, a fluid that discovers the leak (TEKSCENT) and a trained dog. A key component in the fluid is another substance that is emitted in the event of a leak and that rises directly up to the surface of the earth, where the dog can detect it in concentrations as low as one part per quintillion (1 in 10^{18}). Natural gas in itself has no odor, but usually small amounts of ethanethiol are added to it to give the gas a scent. Dogs that sniff at gas pipes can therefore search for this substance and find the location of the leak.[68] The dogs were used on sixty leak searches. They worked under tough winter weather conditions and also searched for leaks from underwater gas pipelines, with a success rate of 100 percent. The research scientists also carried out chemical tests, which only detected two out of nine leaks.[69]

In urban environments, it is particularly important to check whether gas pipelines are intact. In cities, leaks can have large-scale consequences. During the first trials in the beginning of the 2000s, the Swedish research scientists Bjørn Rosén and Lennart Wetterholm at the Hundcampus dog training center in Sweden investigated whether dogs managed to distinguish between leaks in gas pipes where the gas was still flowing out and leaks where gas had flowed out previously. The dogs were unsuccessful.[70] The next year the dogs were trained to give indications for tetrahydrothiophene (THT), a compound that lingers long after a leak has arisen, as opposed to for natural gas itself.[71] THT contains the additive butyric acid, which dogs can detect easily.[72] This poisonous and flammable compound is an additive in natural gas. The research scientists carried out some new trials, and this time the dogs' indications were correct. With a gentle wind, the dogs detected a leak of 20 parts per million from a distance of at least 20 meters, while in a strong wind they detected the leak from at least 45 meters away.[73]

Luna, a six-year-old Siberian husky, discovered a gas leak in the backyard of the house of her owner, Jenny Conarroe, and her family in Mill Woods, Canada. Luna was let out into the yard one evening in June 2011 and came back covered with mud. The owners didn't notice anything that evening, but the next day they found a place in the backyard where Luna had been digging. There was a big hole there and a puddle that was bubbling, which turned out to be a gas leak. A pipe was leaking and Luna had discovered it thanks to her keen sense of smell. Luna saved the family of eight from an extremely dangerous situation.[74]

Gas Pipe Inspector

Pipe corrosion damage is a problem in the gas industry. The Kårstø gas works in Nord-Rogaland, Norway, supplies the transport and treatment of gas and natural gas condensate from important areas on the Norwegian continental shelf. The gas works has kilometer-long pipe coils, where corrosion damage can occur and lead to problems. Moisture can form under the insulation that causes rust and, in the worst case, holes in the pipes. Every year since 2007, Kårstø has spent more than US$33 million to prevent corrosion damage. In collaboration with Gassco (a state operating company responsible for onshore gas works), Rune Fjellanger from the dog training school Fjellanger Hundeskole in Os south of Bergen has used five dogs—hunting springer spaniels and Belgian sheepdogs (Malinois)—to locate corrosion damages since 2010. The dogs have been trained to recognize the scent of corrosion. Up to now, Gassco has spent approximately US$1 million on the project. The samples are gathered with an ejector, which uses compressed air to create a vacuum. The vacuum sucks air samples into a filter, which is then taken to the laboratory. On a scent carousel, the dogs have the chance to sniff the filter samples taken from sites around the pipes with or without rust. Scent samples are also collected from Kollsnes, Sture, and Mongstad in Norway and Emden in Germany. If the sample contains corrosion, the dogs will give an indication with their paws or by sitting down next to the sample. This particular corrosion damage on the pipe can thereby be repaired. So far the dogs have sniffed 3,300 samples and given correct indications in 92–93 percent of the trials.[75]

13

Other Work Tasks for Sniffer Dogs

Scott and Karen Reynolds from Vermontville, Michigan, started Environmental Canine Services in 2009. They use the dogs Sable, a mixed shepherd breed, and Logan, a mixed long-haired collie, to find the source of pollution in water ponds, surface and irrigation water, and recreational areas. It is important to find out who or what is doing the polluting and remove the source. They have trained Sable and Logan to find human feces, which can contain bacteria (total coliforms, *E. coli*, and *Enterococcus*) and cause illness. In recreation areas, the dogs also check whether sewage is leaking from septic tanks in campers.[1]

The Food and Wine Inspector

Truffles and Other Fungi

Dogs can be trained to find several different types of edible mushrooms.[2] Mushroom detection courses for dogs have been held for many years. In the course of two days, dogs can be trained to track down chanterelles, which have a quite complicated scent composition.[3] There is little difference in the abilities of different breeds of dogs when it comes to learning how to track down mushrooms. All dogs from tiny Chihuahuas to huge wolfhounds can be trained to find whatever edible mushroom you might desire[4]—everything from the penny bun mushroom, the gypsy mushroom, and the hedgehog mushroom to the horn of plenty, funnel chanterelle, and the summer truffle.

Most of us associate the truffle with Italy, and the annual sales volume of the truffle industry is approximately US$590 million.[5] The Lagotto Romagnolo is an Italian water dog breed that is a specialist in searching for truffles.[6] The dogs search close to the ground and have no hunting instinct to speak of, so they are not particularly interested in following the trails of wild animals. Other dog breeds such as the Great Pyrenees, golden retriever, Labrador retriever, German shepherd, Australian shepherd, and Welsh springer spaniel can be trained for truffle hunting. Nonetheless, ideally you should have a Lagotto or another kind of retrieving water dog breed.[7] The Lagotto has been used to find truffles for almost 130 years.[8] The truffle industry is also found in France, Croatia, Spain, Portugal, China, Australia, and the United States.[9] In 2007 a truffle weighing 1.5 kilos was sold for US$330,000 to a Chinese casino owner. Since truffles are so valuable, truffle hunters in southern Europe have been known to put out poisoned food to kill competing dogs.[10] This occurs especially during poor truffle years.[11]

Truffles are fungi that grow underground, at depths of 10 to 20 centimeters in calcareous soil. All truffles live in symbiosis with the roots of host trees.[12] Many species grow together with the roots of hardwood trees such as hazel and oak, but also birch and linden.[13] The fungus supplies the tree with minerals and additional water, and the tree provides the fungus with sugary photosynthesis products abundant in energy. The black summer truffle has a nondescript taste that is also overshadowed by a powerful aroma. The scent of truffles has evolved to ensure that they are eaten and spread. The truffle's development takes place underground where they close their spores into a knob/ball.[14] Their powerful odor attracts insects and other animals. It is the compound dimethyl sulfide that gives away the truffle's location.[15] Many mammals dig up these truffles and eat them. Traditionally pigs have been used to find truffles, but the problem with this is that the pig prefers to eat the truffles. Truffle dogs are also interested in eating the truffles they find, but it is easier to train them not to do this than to train a pig. Dogs are also easier to transport than pigs.[16] Dogs have now taken over more and more of the work tasks of the truffle pigs. However, dogs do expect a reward when they find a truffle.[17] In Sweden the first black truffle was found on August 16, 1977. Twenty years would pass before the next one was found. The lucky truffle finder was Christina Wedén in Gotland. Food lovers are willing to pay between US$1,000 and $1,700 per kilo for the black truffle. The white truffle (*trifola d'Alba Madonna*—"Truffle of the White Mother"—from the Piedmont region of Italy) can cost US$5,000 or more per kilo.[18] In Gotland the truffle season lasts from September to December.[19]

Dogs can learn to recognize different scents even while still in the womb and immediately after birth.[20] If you want to use your dog to find truffles (or something else), it is a good idea to start training early. You can give the puppy's mother truffles so the puppy learns this scent at an early age. If the mother is fed truffles, the taste goes directly into her breast milk and then on to her puppies. And if you continue with the feeding after the puppy has been born, you have gotten off to a good start in training your dog to develop a particular fascination with the scent of truffles. Some dog owners also spread truffle essence on the mother dog's teats so the puppy will associate the scent of truffles with something positive.[21] It is important not to feed dogs just any kind of mushroom. Every year there are reports of dogs that are victims of mushroom poisoning. The dog can be fed truffles since they are intended to be eaten and spread by animals. However, other edible mushrooms are not and should not be fed to a dog. Most edible mushrooms are spread by the wind.[22]

Truffle-hunting trials have been held in France since 1969. Here dogs have to find six truffles in a 25 m^2 space as quickly as possible and give indications for their finds with their paw. If the dogs eat the truffles, they are disqualified.[23]

Vegetables and Fruit

Raw or poorly washed vegetables and fruit can contain droppings from rodents or birds, which in turn can cause illness. American research scientist Melissa L. Partyka at the University of California, Davis, and her colleagues trained two dogs in 2014 to detect contamination in lettuce, spinach, coriander, and tomatoes from the droppings of eight different mammals and birds.[24] The dogs succeeded in 96 percent of the attempts to detect droppings on vegetables and fruit in amounts of at least 2.5 grams. The less droppings there were, the more poorly the dogs performed. They managed correct outcomes in 83 percent of the cases for 0.25 gram of droppings, but in only 36 percent of the cases for 0.025 gram, 17 percent for 0.0025 gram, and 7 percent for 0.00025 gram. The amount of droppings was thus highly significant to the dogs' success rate.

Blue-Green Algae

Fish is healthy and tasty, and many people often enjoy a nice meal of salmon. But if the farmed salmon in Norway were to acquire a moldy and earthy flavor due to substances produced by algae in fish farms, sales would plummet dramatically. Dogs can help fish farmers to avoid this predicament[25] by making sure that algal bloom is detected at an early stage so chemicals that eliminate algal bloom can be added to fish farming enclosures and thereby help farmers avoid substantial financial losses.

Catfish consumption in 2001 was 136 million kilograms, valuing US$670 million, so catfish farming is a highly lucrative business.[26] Unfortunately, moldy and earthy odors are the cause of large losses for catfish farmers in the southern United States.[27] This unpleasant odor can lead to consumers eating less of this kind of fish, choosing other fish types, or switching to other meat instead. Among other sources, some natural blue-green algae are responsible for the production of these aroma compounds (2-Methylisoborneol and geosmin), and they are found in most catfish farming ponds. The bloom of these algae types develops rapidly and without warning in catfish ponds in the summertime. This leads to a rapid uptake of the compounds and their subsequent accumulation in fish meat, and thus the fish cannot be harvested until the compounds have been removed. Keeping the catfish in the ponds for long time periods means added expense in conjunction with feeding and the chemicals required to reduce the algae growth, not to mention a lot of extra work. The delayed harvest can also cause further fish fatalities from illnesses, poor water quality, or wild bird predation. It is estimated that this represents costs of US$15–24 million per year.[28] Furthermore, the poison used to combat blue-green algae can also kill the fish.[29]

To avoid the huge economic losses that algal bloom can inflict on the catfish farming industry, dogs were trained in 2004 by the microbiologist Richard A. Shelby and colleagues at the Aquatic Health Research Laboratory in Auburn, Alabama, to detect these algal bloom odorants at an early stage. Three of the dogs that responded best to the training were chosen for testing to see whether they could detect the odorants in a concentration typical of early stage algal bloom. For the lower concentration of 10 nanogram per liter, the average correct response was 37 percent, 43 percent, and 67 percent for the three dogs. Further tests using normal concentrations showed results between 30 and 95 percent, depending upon the test and the dog. The fish ponds that did not contain these compounds were correctly identified in 96 percent of the cases. The odors change quickly, so farmers need quick analyses.[30] Another study done two years later by the same research scientists showed that the four dogs could detect these compounds also in catfish meat. They had correct outcomes in 81 percent of the cases on a training platform.[31]

Climate change can also be the cause of an increase in algae problems in Norwegian fish farms and leads to large losses of farmed fish. Should Norwegian fish farmers also use dogs to check whether there is algal bloom in fish farming enclosures?

Wine

Tainted corks can give wine a flavor reminiscent of the scent of a moldy cellar, a rotten tree, dust, or soil. There are a number of compounds that can cause a moldy aroma, and one of these is the compound 2,4,6-trichloroanisole (TCA).[32] Wine corks are treated with chlorine, which reacts to phenols (and other woodwork), and if the cork is infected by mold, TCA is produced. Very small amounts of TCA on a cork are sufficient to destroy the wine and an entire year's production of wine can be ruined by one single teaspoon of this compound.[33] Wine producer Cliff Lede in Napa, California, trained the bloodhound Miss Louisa Belle in 2006 to detect tainted wine bottles. She can detect bad wine even through the bottle.[34]

The Guard Dog

A pack of dogs living in proximity to humans in former times would undoubtedly have given these communities an advantage over any enemies without dogs, by providing an early warning signal. Warriors who snuck up on a campsite would have been much easier to detect if there were guard dogs around the camp. People who lived without dogs would correspondingly have been easy prey during such nighttime raids. The dog's keen sense of smell and ability to stand guard and defend led to it being given the task of guarding forts and fortresses.[35] Napoleon used dogs as guard dogs, chaining them to the walls of the city of Alexandria so they would sound the alarm in the event of an impending attack. Many people have watchdogs, and dogs have been used to protect important resources for us, from forest plantations, apple orchards, and vegetable farms in need of protection from deer, to golf courses and property in need of protection from Canadian geese.[36]

Dogs have long been used to watch over livestock to ensure that the animals are not taken by predators.[37] They are effective and can give an early warning signal.[38] In addition to their sense of smell, dogs use their eyesight and hearing for this task. If predators come too close to livestock, the dogs become aggressive. In a village in Turkey, Anatolian shepherd dogs have watched over the nomads' livestock since 1800 BC. The best dog was the one who stayed with the herd or flock and did not waste time chasing animals. It was bred over the course of many generations to protect livestock and not to attack farm animals.[39] In Namibia, the rare and endangered cheetah kills large numbers of livestock when the herd or flock doesn't include a dog as a family member. Before the dog Flintis came to his farm in the late 1990s, Johann Coetzee lost many animals. Cheetahs had killed forty-two sheep in one night, and a few nights later they killed another twenty-nine. In consequence, the owner of the farm also killed many cheetahs to protect his flock, but after Flintis arrived at the farm, cheetahs were no longer a threat since they are afraid of dogs. If the dog sees a predator, it barks furiously to warn the intruder that it has been discovered and that it will attack should the predator come any closer. Flintis worked around the clock and protected both the sheep and the cheetahs.[40]

Sheep husbandry is an important industry in many places. Oddly enough, guard dogs are not used. Dogs would not only be able to help the sheep but also their natural predators. We often hear of farmers who have lost sheep to wolves, bears, wolverines, or lynx. And we read all the time about how another one of these predators has been shot. Had we included dogs in the flock, the lives of many predators would very likely have been saved.

Detection of the Scent of Cows in Estrus and Pregnant Polar Bears

It is important to discover when milking cows are ready for mating. In 1978 dogs were already being used for this purpose in the United States.[41] In 2011 the research scientists Carola Fischer-Tenhagen at Freie University in Berlin and dog trainer Lennart Wetterholm from the HundCampus dog training center in Sweden trained seven dogs to detect when cows are in heat.[42] They investigated whether dogs could differentiate between urine, vaginal secretion, and milk from cows that were and were not in estrus. They found that the dogs detected the scent of estrus in urine and vaginal secretions with more than 80 percent accuracy. Many farms keep guard dogs, and it would be a good idea to train these dogs to detect when the cows are in estrus, saving farmers valuable time that they can use for other work tasks. The experiment using milk was not equally successful. The first part of the training went well, but in the second round the dogs were so distracted by the scent of milk as a potential food reward that they no longer showed any interest in searching for the scent of estrus.[43] Dogs were also trained by Fischer-Tenhagen and her colleagues in 2011 to detect estrus scent compounds in the saliva of cows, and the success rate varied from 40 to 75 percent (an average of 58 percent). Although the dogs were less successful here than with vaginal secretion, this trial showed at least that dogs are able to identify the period of estrus by sniffing at the mouth and nose of the cow.[44]

Due to climate change, the polar bear is now an endangered species.[45] Polar bears are being bred in captivity. In Toledo, Ohio, two polar bear cubs were born in captivity in 2013. Only one more cub was born in captivity in the entire United States. It is therefore important to determine mating success—in other words, whether or not the female has been impregnated. It is not easy to establish whether female polar bears are pregnant, and false pregnancies are common. The dog trainer Matt Skogen from Iron Heart High Performance Working Dogs at the Cincinnati Zoo was summoned to lend a hand with this task. After mating in a zoological garden, the female polar bear is separated from the male and goes into hibernation for the winter. The research scientists set up a video camera to monitor if and when the polar bear gives birth in the course of the winter. It would make more sense to establish whether the female polar bear was pregnant in the first place and before she took her winter nap. In 2013 a beagle named Elvis was thus given the job of figuring this out. There are specific proteins found only in the feces of pregnant females, and these are what Elvis sniffs out. He checked twenty-two droppings from female polar bears in fourteen different zoos in the United States and Canada, and achieved correct results in 97 percent of the cases.[46]

Ore and Minerals Detector

Ore is a type of rock that contains a lot of minerals and from which metals can be extracted. All the metal we use in daily life has been extracted from ore minerals. In the late 1960s, dogs were used to find different types of ore such as pyrite, pyrrhotite, chalcopyrite, and arsenopryte.[47] Dogs have later been used to find zinc, nickel, and copper as well. It is claimed that dogs can find minerals located up to 12–15 meters underground.[48] In Sweden Peter Bergman's company OreDog is investigating whether dogs are able to distinguish between arsenopryte (an indicator of gold) and the highly similar pyrrhotite (which does not contain gold). He will also test whether or not dogs can find rare earth metals such as yttrium, cerium, terbium, and lanthanum. These are important minerals used in batteries, wind-power stations, hard disk drives, and fiber optic cables. In the United States, dogs are used to find gold, silver, and brass.[49]

Contaminants and Toxins Detector

Dogs are used to find PCB (polychlorinated biphenyl), PAH (polycyclic aromatic hydrocarbons), and mercury.[50] The use of PCB is now prohibited many places. PCB was used in the glue of double-glazed windowpanes, capacitors, lighting fixtures, concrete, grouting cement, paint, and other building construction materials in the 1930s. The toxin is not dangerous as long as it stays in the building, but once the materials in the building start to break down and PCB gets into the environment, it can be taken up by fish, birds, and mammals.[51]

In 1997 research scientist Allison Crook from Australia started training the dog Norm as the first in the world to detect organic environmental toxins (organochlorines) in the soil, especially dieldrin, aldrin, DDT (dichlorodiphenyltrichloroethane) and its degradation products DDD (dichlorodiphenyldichloroethane) and DDE (dichlorodiphenyldichloroethane). Chemical remains of organochlorines have been found in steak from Australia. To find different types of contaminants on a farm, the dog Norm was trained to detect very low concentrations of aldrin, dieldrin, and DDT (1 part per million or less) in the soil. The use of dogs for this purpose saves time and reduces the number of soil samples necessary to find the contaminated areas.[52]

Dogs also search for other toxins (toluene, 2,4,5- and 2,4,6-trichlorophenol, and 1,2,3-trichloropropane) to improve safety standards for humans.[53] One of the most poisonous substances in the world is mercury, which is liquid at room temperature. Mercury is found in thermometers, barometers, and lightbulbs. Mercury detection is no easy task. It can be found in a school chemistry lab, in a hospital, or in an old factory building.[54] The inhalation of mercury vapor is hazardous to the health and can cause brain damage.[55] It also has serious consequences for fish and mammals.[56] In 1999 two dogs inspected 1,100 schools in Sweden for mercury and found approximately 1.4 tons![57]

Turid Buvik and Per Johan Brandvik have been training the dogs Jippi (a border collie) and Tara (a long-haired dachshund) in oil pollution detection. The dogs detect oil spillage covered with snow, ice, or littoral sediments. The dogs work effectively in temperatures down to −20°C and with strong winds for long periods of time. The oil samples (0.4 liter) that were buried deep in the ice of the fjord (30 cm) were put out a week before the dogs arrived at the island Svalbard, Norway. The dogs located the samples with great precision and were able to determine the size of the oil spillage area. The dogs also discovered evidence of large-scale pollution stemming from leaks in a 400-liter oil drum found on the ice. They detected the pollution from a distance of 5 kilometers. Later the dogs were trained to ignore small amounts of oil and to focus on larger spills, so cleaning up these areas could be prioritized. An effective way of using dogs in the future would be to have them identify oil pollution in coastal regions.[58] Unfortunately, far too often we hear stories of ships that have run aground, causing oil cargo to spill out into the ocean. Oil spills have a hugely negative impact on animal life. Dogs trained to detect oil will also be able to detect creosote, which is predominantly made up of polycyclic aromatic hydrocarbons and is a generic term for the products wood-tar creosote and coal-tar creosote.

Dogs can also be trained to detect contaminated carcasses and to find animals that have died of poisoning.[59] In Spain hunting is very popular, which sometimes culminates in conflicts between humans and predatory animals. Using poison to kill predatory animals is illegal, but some hunters and gamekeepers nonetheless put out poisoned bait to kill foxes, feral cats and dogs, and other predators.[60]

Dogs have also been used to detect hazardous chemical leaks from bulldozers, tractor diggers, and power shovels.[61]

Stolen World Cup Trophy

On March 20, 1966, the FIFA World Cup trophy was stolen in London just four months before the World Cup final in football was to be played. A week later, in South London, David Corbett's mixed-breed collie Pickles picked up the scent of something in a yard. Under a hedge lay an object wrapped up in a newspaper. Pickles had found the missing World Cup trophy and became a national hero overnight.[62]

Golf Balls

Goya, a five-year-old mixed breed (Labrador and golden retriever), sniffs out golf balls together with the mistress of the house, Elisabeth Thomsen. Goya's record is 27 golf balls in one day. In September and October 2012, and from May 1 to September 16, 2013, she sniffed out as many as 1,053 golf balls. She finds the golf balls even if they are lying in muddy water and she can't see them. In the course of fourteen days in the early summer of 2014, she sniffed her way to no less than 123 golf balls. They were lying under snow, heather or grass, or on the bottom of brooks and bogs. Nevertheless, she finds them after a lot of digging![63] The world record, however, may belong to the cocker spaniel Rikki, who during the Second World War found 40,000 golf balls in the area around Saunton Golf Club in southwest England.[64]

ACKNOWLEDGMENTS

I would like to extend my thanks to all those who have contributed hints and advice, those who have accepted me as a visitor, and those who have read and made comments on the contents of the book. A special thank you goes to Ole K. Auten (chapters 7 and 9), Terje Groth Berntsen (chapters 1 and 2), Dag K. Bjerketvedt (chapters 10–12), Susan Bulanda (chapters 5 and 6), Turid Buvik (chapter 10), Olav Inge Edvardsen (chapter 11), Birgit Espedalen (chapters 10–12), Rune Fjellanger (chapters 1, 9, and 12), Asbjørn Grande (chapters 7–9), Claire Guest (chapter 10), Monica Hagerup (chapter 5), Inger Hanssen-Bauer (chapter 6), Anne Hermansen (chapters 1–3, 7, and 8), György Horvath (chapter 10), Tor Iljar (chapter 12), Kai Iversen (chapter 7), Per Tore Iversen (chapter 5), Tadeusz Jezierski (chapter 7), Imke Jürgens (chapters 10–12), Jon Einar Karlsen (chapter 7), Kristin Killingmo (chapter 13), Torun Knapperholen (chapter 8), Marcia Koenig (chapter 7), Rolf von Krogh (chapter 8), Anne Molia (chapter 13), Howard Parker (chapter 6), Kåre Vidar Pedersen (chapter 6), Paola A. Prada (chapter 3), Ole Reitan (chapters 3–6 and

11), Christian A. Robstad (chapter 11), Gary S. Settles (chapter 2), Knut Skår (chapter 5), Cato Sletten (chapter 12), Siri Stedje (Chapter 10), Bjørn G. Steen (the entire book), Øyvind Steifetten (chapters 10 and 11), Monica Alterskjær Sundset (chapter 1), Thor Svendsen (chapter 8), Per Arne Sødal (chapter 7), Mona Sæbø (chapter 1), Torun Thomassen (chapters 1–3), Helga Veronica Tinnesand (chapters 10 and 11), Silje Vang (chapter 6), Vidar Vestreng (chapter 8), Andreas Zedrosser (chapters 10 and 11), and Frode Ødegaard (chapter 12).

I would also like to thank the staff of the library at Telemark University College in Bø, Norway, for procuring countless books, reports, and scientific papers about dogs, and to Dean Tone Jøran Oredalen, who has supported my dog project from the start. A special thanks also goes to Tor Iljar at Dogpoint for a very educational dog project and to my PhD candidate Hannah B. Cross, to master's degree student Christin Beate Johnsen, and dog enthusiast Beate Jaspers for their fantastic efforts during all our dog training sessions. A special thank you to the editors Bjørn Olav Jahr at Gyldendal Sakprosa and Christie Henry and Erin DeWitt at the University Chicago Press for many constructive comments and their great efforts toward the publication of this book.

And finally: Frid Elisabeth Berge and Yrja Skjærum are to be thanked for their patience and support while I was working on this book.

It is my hope that reading this book will offer you a closer look into the dog's unique olfactory universe.

NOTES

CHAPTER ONE

1. G. Salisbury and L. Salisbury, *The Cruelest Miles: The Heroic Story of Dogs and Men in a Race against an Epidemic* (W. W. Norton & Company, 2003), 223–25.
2. A. Miklosi, *Dog Behaviour, Evolution, and Cognition* (Oxford University Press, 2007).
3. L. R. Mehrkam and C. D. L. Wynne, "Behavioral Differences among Breeds of Domestic Dogs (*Canis lupus familiaris*): Current Status of the Science," *Applied Animal Behaviour Science* 155 (June 2014): 12–27, http://dx.doi.org/10.1016/j.applanim.2014.03.005; S. Robin et al., "Genetic Diversity of Canine Olfactory Receptors," *BMC Genomics* 10, no. 21 (2009); K. Lord, L. Coppinger, and R. Coppinger, in *Genetics and Behaviour of Domestic Animals*, ed. T. Grandin and M. J. Deesing, 2nd ed. (Academic Press, 2014).
4. M. E. Gompper, "The Dog-Human-Wildlife Interface: Assessing the Scope of the Problem," in *Free-Ranging Dogs and Wildlife Conservation*, ed. M. E. Gompper (Oxford University Press, 2014), 9–54.
5. K. M. C. Cline, "Psychological Effects of Dog Ownership: Role Strain, Role Enhancement, and Depression," *Journal of Social Psychology* 150 (2010): 117–31; R. Austin, *How to Train Your Dog to Find Gold* (Golddogs, 2008).
6. W. S. Helton, ed., *Canine Ergonomics: The Science of Working Dogs* (CRC Press, 2009).

7. R. T. McGowan et al., "Positive Affect and Learning: Exploring the 'Eureka Effect' in Dogs," *Animal Cognition* 17 (2014): 577–87, doi:10.1007/s10071-013-0688-x.
8. R. K. Wayne, "Molecular Evolution of the Dog Family," *Trends Genetics* 9 (1993): 218–24.
9. B. Hare and V. Woods, *The Genius of Dogs: How Dogs Are Smarter than You Think* (Dutton, 2013).
10. M. Hindrikson et al., "Bucking the Trend in Wolf-Dog Hybridization: First Evidence from Europe of Hybridization between Female Dogs and Male Wolves," *PLOS ONE* 7 (2012), doi:10.1371/journal.pone.0046465.
11. C. Vila et al., "Multiple and Ancient Origins of the Domestic Dog," *Science* 276 (1997): 1687–89, doi:10.1126/science. 276.5319.1687; A. H. Freedman et al., "Genome Sequencing Highlights Genes under Selection and the Dynamic Early History of Dogs," http://arxiv.org/abs/1305.7390; O. Thalmann et al., "Complete Mitochondrial Genomes of Ancient Canids Suggest a European Origin of Domestic Dogs," *Science* 342 (2013): 871–74, doi:10.1126/science.1243650; A. H. Freedman et al., "Genome Sequencing Highlights the Dynamic Early History of Dogs," *PLoS Genetics* 10 (2014), e1004016, doi:10.1371/journal.pgen.1004016.
12. J. Clutton-Brock, "Origins of the Dog: Domestication and Early History," in *The Domestic Dog: Its Evolution, Behaviour and Interaction with People*, ed. J. Serpell (Cambridge University Press, 1995), 7–20.
13. N. D. Ovodov et al., "A 33,000-Year-Old Incipient Dog from the Altai Mountains of Siberia: Evidence of the Earliest Domestication Disrupted by the Last Glacial Maximum," *PLOS ONE* 6 (2011), https://doi.org/10.1371/journal.pone.0022821.
14. Thalmann et al., "Complete Mitochondrial Genomes of Ancient Canids."
15. E. Axelsson et al., "The Genomic Signature of Dog Domestication Reveals Adaptation to a Starch-Rich Diet," *Nature* 495 (2013): 360–64, doi:10.1038/nature11837.
16. Ibid.
17. A. Bhadra and A. Bhadra, "Preference for Meat Is Not Innate in Dogs," *Journal of Ethology* 32 (2014): 15–22, doi:10.1007/s10164-013-0388-7.
18. B. G. Dias and K. J. Ressler, "Parental Olfactory Experience Influences Behavior and Neural Structure in Subsequent Generations," *Nature Neuroscience* 17 (2014): 89–96.
19. Freedman et al., "Genome Sequencing Highlights the Dynamic Early History of Dogs."
20. R. Bonanni et al., "Effect of Affiliative and Agonistic Relationships on Leadership Behaviour in Free-Ranging Dogs," *Animal Behaviour* 79 (2010): 981–91, doi:10.1016/j. anbehav.2010.02.021.
21. N. J. Rooney and J. W. S. Bradshaw, "An Experimental Study of the Effects of Play upon the Dog-Human Relationships," *Applied Animal Behaviour Science* 75 (2002): 161–76, doi:10.1016/s0168-1591(01)00192-7.
22. C. Ritter, Z. Viranyi, and F. Range, "Who Is More Tolerant? Cofeeding in Pairs of Pack-Living Dogs (*Canis familiaris*) and wolves (*Canis lupus*)," Third Canine Science Forum (Barcelona, July 25–27, 2012).

23. A. Goldblatt, I. Gazit, and J. Terkel, "Olfaction and Explosives Detector Dogs," in *Canine Ergonomics: The Science of Working Dogs*, ed. W. S. Helton (CRC Press, 2009), 135–74.
24. K. Mehus-Roe, *Working Dogs: True Stories of Dogs and Their Handlers* (BowTie Press, 2003); M. C. Roy and C. Villeneuve, *Canine Angels* (Roy and Newtown, 2011); L. Rogak, *Dogs of Courage: The Heroism and Heart of Working Dogs around the World* (Tomas Dunne Books, St. Martin's Griffin Press, 2012).
25. Helton, ed., *Canine Ergonomics*.
26. Hare and Woods, *The Genius of Dogs*.
27. J. Kaminski, J. Call, and J. Fischer, "Word Learning in a Domestic Dog: Evidence for 'Fast Mapping,'" *Science* 304 (2004): 1682–83, doi:10.1126/science.1097859.
28. J. W. Pilley and A. K. Reid, "Border Collie Comprehends Object Names as Verbal Referents," *Behavioural Processes* 86 (2011): 184–95, doi:10.1016/j.beproc.2010.11.007.
29. J. W. Pilley, "Border Collie Comprehends Sentences Containing a Prepositional Object, Verb, and Direct Object," *Learning and Motivation* 44 (2013): 229–40, doi:10.1016/j.lmot.2013.02.003.
30. Pilley and Reid, "Border Collie Comprehends Object Names."
31. Pilley, "Border Collie Comprehends Sentences"; J. W. Pilley and H. Hinzmann, *Chaser: Unlocking the Genius of the Dog Who Knows a Thousand Words* (Houghton Mifflin Harcourt, 2013); Hare and Woods, *The Genius of Dogs*.
32. A. Cook, J. Arter, and L. F. Jacobs, "My Owner, Right or Wrong: The Effect of Familiarity on the Domestic Dog's Behavior in a Food-Choice Task," *Animal Cognition* 17 (2014): 461–70, doi:10.1007/s10071-013-0677-0.
33. Hare and Woods, *The Genius of Dogs*.
34. J. Brauer, K. Schonefeld, and J. Call, "When Do Dogs Help Humans?" *Applied Animal Behaviour Science* 148 (2013): 138–49, doi:10.1016/j.applanim.2013.07.009.
35. Hare and Woods, *The Genius of Dogs*.
36. G. J. Adams and K. G. Johnson, "Sleep, Work, and the Effects of Shift Work in Drug Detector Dogs *Canis familiaris*," *Applied Animal Behaviour Science* 41 (1994): 115–26, doi:10.1016/0168-1591(94)90056-6.
37. P. MacKay et al., "Scat Detection Dogs," in *Noninvasive Survey Methods for Carnivores*, ed. R. A. Long et al. (Island Press, 2008), 183–222.
38. M. Nau, *Snooping Around! Train Your Dog to Be an Expert Sniffer* (Cadmos Publishing, 2011).
39. National Association of Canine Scent Work, https://www.nacsw.net/; K9 Nose Work, http://www.k9nosework.com/; DogPoint, www.dogpoint.no/.
40. Spesialsøk, http://www.spesialsok.no/; Nau, *Snooping Around!*
41. A. L. Kvam, *Nesearbeid for hund: Skritt for skritt fra godbitsok til sporsok* (Tun forlag AS, 2005); D. Johnen, W. Heuwieser, and C. Fischer-Tenhagen, "Canine Scent Detection—Fact or Fiction?" *Applied Animal Behaviour Science* 148 (2013): 201–8.
42. A. L. Kvam and T. Rugaas, in *Nose Work: Search Games* (DVD, Haqihana, Torino, Italy, 2012).

43. Fjellanger Hundesenter, http://www.fjellanger.net/artikler/88-hunden-ilaboratoriet.
44. J. E. King, R. F. Becker, and J. E. Markee, "Studies on Olfaction Discrimination in Dogs: (3) Ability to Detect Human Odour Trace," *Animal Behaviour* 12 (1964): 311–15.
45. D. G. Moulton, E. H. Ashton, and J. T. Eayrs, "Studies in Olfactory Acuity: 4. Relative Detectability of N-Alipatic Acids by the Dog," *Animal Behaviour* 8 (1960): 129–33; D. G. Moulton and D. A. Marshall, "The Performance of Dogs in Detecting α-Ionone in the Vapor Phase," *Journal of Comparative Physiology* 110 (1976): 287–306; Fjellanger Hundesenter, http://www.fjellanger.net/artikler/88-hunden-ilaboratoriet.
46. F. Rosell and L. J. Sundsdal, "Odorant Source Used in Eurasian Beaver Territory Marking," *Journal of Chemical Ecology* 27 (2001): 2471–91, doi:10.1023/a:1013627515232.
47. F. Rosell and F. Bergan, "Free-Ranging Eurasian Beavers, Castor fiber, Deposit Anal Gland Secretion When Scent Marking," *Canadian Field-Naturalist* 112 (1998): 532–35.
48. Rosell and Sundsdal, "Odorant Source Used in Eurasian Beaver Territory Marking."
49. L. Horn, L. Huber, and F. Range, "The Importance of the Secure Base Effect for Domestic Dogs—Evidence from a Manipulative Problem-Solving Task," *PLOS ONE* 8 (2013), doi:10.1371/journal.pone.0065296.
50. I. Merola et al., "Dogs' Comprehension of Referential Emotional Expressions: Familiar People and Familiar Emotions Are Easier," *Animal Cognition* 17 (2014): 373–85, doi:10.1007/s10071-013-0668-1.
51. S. Bulanda with L. Bulanda, *Ready! Training the Search and Rescue Dog*, 2nd ed. (Kennel Club Books, 2010).
52. C. Millan and M. J. Peltier, *Cesar's Way: The Natural, Everyday Guide to Understanding and Correcting Common Dog Problems* (Hodder and Stoughton, 2012).
53. C. M. Browne et al., "What Dog Owners Read: A Review of Best-Selling Books," Third Canine Science Forum (Barcelona, July 25–27, 2012).
54. C. Fugazza and A. Miklosi, "Should Old Dog Trainers Learn New Tricks? The Efficiency of the Do As I Do Method and Shaping/Clicker Training Method to Train Dogs," *Applied Animal Behaviour Science* 153 (2014): 53–61, http://dx.doi.org/10.1016/j.applanim.2014.01.009.
55. Goldblatt, Gazit, and Terkel, "Olfaction and Explosives Detector Dogs."
56. K. G. Furton and L. J. Myers, "The Scientific Foundation and Efficacy of the Use of Canines as Chemical Detectors for Explosives," *Talanta* 54 (2001): 487–500, doi:10.1016/s0039-9140(00)00546-4.
57. O. Pfungst, *Clever Hans (The Horse of Mr. Von Osten): A Contribution to Experimental Animal and Human Psychology*, trans. C. L. Rahn (Henry Holt, 1911).
58. G. A. A. Schoon and R. Haak, *K9 Suspect Discrimination* (Detseling Enterprises Ltd., 2002).

59. V. Szetei et al., "When Dogs Seem to Lose Their Nose: An Investigation on the Use of Visual and Olfactory Cues in Communicative Context between Dog and Owner," *Applied Animal Behaviour Science* 83 (2003): 141–52, doi:10.1016/s0168-1591(03)00114-x.
60. L. Lit, J. B. Schweitzer, and A. M. Oberbauer, "Handler Beliefs Affect Scent Detection Dog Outcomes," *Animal Cognition* 14 (2011): 387–94, doi:10.1007/s10071-010-0373-2.
61. K. Soproni et al., "Comprehension of Human Communicative Signs in Pet Dogs (*Canis familiaris*)," *Journal of Comparative Psychology* 115 (2001): 122–26, doi:10.1037//0735-7036.115.2.122; Z. Viranyi et al., "Dogs Respond Appropriately to Cues of Humans' Attentional Focus," *Behavioural Processes* 66 (2004): 161–72, doi:10.1016/j.beproc.2004.01.012; C. Schwab and L. Huber, "Obey or Not Obey? Dogs (*Canis familiaris*) Behave Differently in Response to Attentional States of Their Owners," *Journal of Comparative Psychology* 120 (2006): 169–75, doi:10.1037/0735-7036.120.3.169.
62. A. Schoon et al., "Detecting Corrosion under Insulation (CUI) Using Trained Dogs: A Novel Approach," *Materials Evaluation* 72 (2014): 142–48.
63. A. A. Kasparson, J. Badridze and V. V. Maximov, "Colour Cues Proved to Be More Informative for Dogs than Brightness," *Proceedings of the Royal Society B: Biological Sciences* 280 (2013), doi:10.1098/rspb.2013.1356.
64. J. Neitz, T. Geist, and G. H. Jacobs, "Color Vision in the Dog," *Visual Neuroscience* 3 (1989): 119–25, doi:10.1017/S0952523800004430.
65. R. H. Douglas and G. Jeffery, "The Spectral Transmission of Ocular Media Suggests Ultraviolet Sensitivity Is Widespread among Mammals," *Proceedings of the Royal Society B: Biological Sciences* 281 (2014), doi:10.1098/rspb.2013.2995.
66. Johnen, Heuwieser, and Fischer-Tenhagen, "Canine Scent Detection—Fact or Fiction?"
67. L. Falt et al., *Sparhunden och lukterna* (SWDI forlag, 2011).
68. S. Gadbois and C. Reeve, "Canine Olfaction: Scent, Sign, and Situation," in *Domestic Dog Cognition and Behavior*, ed. A. Horowitz, (Springer Berlin Heidelberg, 2014), 3–29.
69. Johnen, Heuwieser, and Fischer-Tenhagen, "Canine Scent Detection—Fact or Fiction?"
70. F. Rosell et al., "Ecological Impact of Beavers *Castor fiber* and *Castor canadensis* and Their Ability to Modify Ecosystems," *Mammal Review* 35 (2005): 248–76, doi:10.1111/j.1365-2907.2005.00067.x; F. Rosell, "The Function of Scent Marking in Beaver (*Castor fiber*) Territorial Defence" (PhD diss., Norwegian University of Science and Technology, 2002); R. McEwing et al., "A DNA Assay for Rapid Discrimination between Beaver Species as a Tool for Alien Species Management," *European Journal of Wildlife Research* 60 (2014): 547–50, doi:10.1007/s10344-014-0803-6.
71. H. Parker et al., "Invasive North American Beaver *Castor canadensis* in Eurasia: A Review of Potential Consequences and a Strategy for Eradication," *Wildlife Biology* 18 (2012): 354–65, doi:10.2981/12-007.

72. McGowan et al., "Positive Affect and Learning."
73. Ibid.
74. A. Quaranta, M. Siniscalchi and G. Vallortigara, "Asymmetric Tail-Wagging Responses by Dogs to Different Emotive Stimuli," *Current Biology* 17 (2007): R199–R201, doi:10.1016/j.cub.2007.02.008.
75. N. J. Rooney, J. W. S. Bradshaw, and H. Almey, "Attributes of Specialist Search Dogs—A Questionnaire Survey of UK Dog Handlers and Trainers," *Journal of Forensic Sciences* 49 (2004): 300–306.
76. N. J. Rooney and J. W. S. Bradshaw, "Breed and Sex Differences in the Behavioural Attributes of Specialist Search Dogs—A Questionnaire Survey of Trainers and Handlers," *Applied Animal Behaviour Science* 86 (2004): 123–35, doi:10.1016/j. applanim.2003.12.007.
77. L. M. Tomkins, P. C. Thomson, and P. D. McGreevy, "Associations between Motor, Sensory and Structural Lateralisation and Guide Dog Success," *Veterinary Journal* 192 (2012): 359–67, doi:10.1016/j. tvjl.2011.09.010; L. M. Tomkins et al., "Lateralization in the Domestic Dog (*Canis familiaris*): Relationships between Structural, Motor, and Sensory Laterality," *Journal of Veterinary Behavior—Clinical Applications and Research* 7 (2012): 70–79, doi:10.1016/j.jveb.2011.07.001.
78. S. D. Gosling, V. S. Y. Kwan, and O. P. John, "A Dog's Got Personality: A Cross-Species Comparative Approach to Personality Judgments in Dogs and Humans," *Journal of Personality and Social Psychology* 85 (2003): 1161–69, doi:10.1037/0022-3514 .85.6.1161; M. Maejima et al., "Traits and Genotypes May Predict the Successful Training of Drug Detection Dogs," *Applied Animal Behaviour Science* 107 (2007): 287–98, doi:10.1016/j.applanim.2006.10.005; P. Jensen, *Hundens sprak och tankar* (Natur & Kultur, 2012); A. V. Kukekova, L. N. Trut, and G. M. Acland, "Genetics of Domesticated Behaviour in Dogs and Foxes," in *Genetics and Behaviour of Domestic Animals*, ed. T. Grandin and M. J. Deesing, 2nd ed. (Academic Press, 2014).
79. K. Svartberg, "A Comparison of Behaviour in Test and in Everyday Life: Evidence of Three Consistent Boldness-Related Personality Traits in Dogs," *Applied Animal Behaviour Science* 91 (2005): 103–28, doi:10.1016/j.applanim.2004.08.030; K. Svartberg, "Breed-Typical Behaviour in Dogs—Historical Remnants or Recent Constructs?" *Applied Animal Behaviour Science* 96 (2006): 293–313, doi:10.1016/j. applanim.2005.06.014.
80. J. Ley, P. Bennett, and G. Coleman, "Personality Dimensions that Emerge in Companion Canines," *Applied Animal Behaviour Science* 110 (2008): 305–17, doi:10.1016/ j.applanim.2007.04.016; K. M. Mornement et al., "Development of the Behavioural Assessment for Re-Homing K9's (B.A.R.K.) Protocol," *Applied Animal Behaviour Science* 151 (2014): 75–83, http://dx.doi.org/10.1016/j.applanim.2013.11.008.
81. E. Mirko, A. Doka, and A. Miklosi, "Association between Subjective Rating and Behaviour Coding and the Role of Experience in Making Video Assessments on the Personality of the Domestic Dog (*Canis familiaris*)," *Applied Animal Behaviour Science* 149 (2013): 45–54.

82. P. Foyer et al., "Early Experiences Modulate Stress Coping in a Population of German Shepherd Dogs," *Applied Animal Behaviour Science* 146 (2013): 79–87, doi:10.1016/j.applanim.2013.03.013.
83. Svartberg, "A Comparison of Behaviour in Test and in Everyday Life"; P. Foyer et al., "Early Experiences Modulate Stress Coping in a Population of German Shepherd Dogs"; I. Svobodova et al., "Testing German Shepherd Puppies to Assess Their Chances of Certification," *Applied Animal Behaviour Science* 113 (2008): 139–49, doi:10.1016/j.applanim.2007.09.010.
84. C. L. Battaglia, "Periods of Early Development and the Effects of Stimulation and Social Experiences in the Canine," *Journal of Veterinary Behavior-Clinical Applications and Research* 4 (2009): 203–10, doi:10.1016/j.jveb.2009.03.003.
85. P. Řezáč et al., "Factors Affecting Dog-Dog Interactions on Walks with Their Owners," *Applied Animal Behaviour Science* 134 (2011): 170–76, doi:10.1016/j.applanim.2011.08.006.
86. M. Wedl et al., "Relational Factors Affecting Dog Social Attraction to Human Partners," *Interaction Studies* 11 (2010): 482–503, doi:10.1075/is.11.3.09wed.
87. P. Řezáč et al., "Factors Affecting Dog-Dog Interactions on Walks with Their Owners."
88. K. Kotrschal et al., "Dyadic Relationships and Operational Performance of Male and Female Owners and Their Male Dogs," *Behavioural Processes* 81 (2009): 383–91, doi:10.1016/j.beproc.2009.04.001.
89. S. Riemer et al., "Choice of Conflict Resolution Strategy Is Linked to Sociability in Dog Puppies," *Applied Animal Behaviour Science* 149 (2013): 36–44, http://dx.doi.org/10.1016/j. applanim.2013.09.006.
90. Quaranta, Siniscalchi, and Vallortigara, "Asymmetric Tail-Wagging Responses by Dogs to Different Emotive Stimuli"; S. Coren, "What a Wagging Dog Tail Really Means: New Scientific Data," *Psychology Today*, December 5, 2011, https://www.psychologytoday.com/blog/canine-corner/201112/what-wagging-dog-tail-really-means-new-scientific-data.
91. M. Siniscalchi et al., "Seeing Left- or Right-Asymmetric Tail Wagging Produces Different Emotional Responses in Dogs," *Current Biology* 23 (2013): 2279–82.
92. https://en.wikipedia.org/wiki/List_of_most_popular_dog_breeds.

CHAPTER TWO

1. "Thomassens hundepensjonat og treningssenter," http://kennelometyst.dinstudio.no/2/7/thomassens-hundepensjonat-og-treningssenter/.
2. T. Thomassen, e-mail message to author, November 2013).
3. W. G. Syrotuck, *Scent and the Scenting Dog* (Arner Publishing, 2000).
4. C. M. Blatt, C. R. Taylor, and M. B. Habal, "Thermal Panting in Dogs: Lateral Nasal Gland, a Source of Water for Evaporative Cooling," *Science* 177 (1972): 804–5, doi:10.1126/science.177.4051.804.

5. K. Lord, "A Comparison of the Sensory Development of Wolves (*Canis lupus lupus*) and Dogs (*Canis lupus familiaris*)," *Ethology* 119 (2013): 110–20, doi:10.1111/eth.12044.
6. D. L. Wells and P. G. Hepper, "Prenatal Olfactory Learning in the Domestic Dog," *Animal Behaviour* 72 (2006): 681–86, doi:10.1016/j.anbehav.2005.12.008.
7. P. G. Hepper and D. L. Wells, "Perinatal Olfactory Learning in the Domestic Dog," *Chemical Senses* 31 (2006): 207–12, doi:10.1093/chemse/bjj020.
8. Ibid.; Wells and Hepper, "Prenatal Olfactory Learning in the Domestic Dog."
9. H. E. Evans and A. de Lahunta, eds., "The Respiratory System," in *Miller's Anatomy of the Dog*, 4th ed. (Elsevier Saunders, 2013), 463–93.
10. G. S. Settles, e-mail message to author, September 15, 2013.
11. B. A. Craven, E. G. Paterson, and G. S. Settles, "The Fluid Dynamics of Canine Olfaction: Unique Nasal Airflow Patterns as an Explanation of Macrosmia," *Journal of the Royal Society Interface* 7 (2010): 933–43, doi:10.1098/rsif.2009.0490.
12. Ibid.
13. T. Eiting et al., "The Role of the Olfactory Recess in Olfactory Airflow," *Journal of Experimental Biology* 217 (2014): 1799–803.
14. B. A. Craven et al., "Reconstruction and Morphometric Analysis of the Nasal Airway of the Dog (*Canis familiaris*) and Implications Regarding Olfactory Airflow," *Anatomical Record: Advances in Integrative Anatomy and Evolutionary Biology* 290 (2007): 1325–40, doi:10.1002/ar.20592; Craven, Paterson, and Settles, "The Fluid Dynamics of Canine Olfaction."
15. G. S. Settles, e-mail message to author, September 15, 2013.
16. Craven et al., "Reconstruction and Morphometric Analysis of the Nasal Airway of the Dog."
17. M. J. Lawson et al., "A Computational Study of Odorant Transport and Deposition in the Canine Nasal Cavity: Implications for Olfaction," *Chemical Senses* 37 (2012): 553–66, doi:10.1093/chemse/bjs039.
18. J. M. Henshaw, *A Tour of the Senses: How Your Brain Interprets the World* (Johns Hopkins University Press, 2012).
19. Craven et al., "Reconstruction and Morphometric Analysis of the Nasal Airway of the Dog."
20. Henshaw, *A Tour of the Senses.*
21. R. Gerritsen and R. Haak, *K9 Professional Tracking: A Complete Manual for Theory and Training* (Detselig Enterprises Ltd., 2001); P. Quignon et al., "Genetics of Canine Olfaction and Receptor Diversity," *Mammalian Genome* 23 (2012): 132–43, doi:10.1007/s00335-011-9371-1.
22. P. Quignon, S. Robin, and F. Galibert, "Canine Olfactory Genetics," in *The Genetics of the Dog*, ed. E. Ostrander and A. Ruvinsky, 2nd ed. (CABI, 2012), 375–93.
23. Y. Gilad and D. Lancet, "Population Differences in the Human Functional Olfactory Repertoire," *Molecular Biology and Evolution* 20 (2003): 307–14, doi:10.1093/molbev/msg013.

24. Quignon, Robin, and Galibert, "Canine Olfactory Genetics."
25. S. S. Steiger et al., "Avian Olfactory Receptor Gene Repertoires: Evidence for a Well-Developed Sense of Smell in Birds?" *Proceedings of the Royal Society B: Biological Sciences* 275 (2008): 2309–17, doi:10.1098/rspb.2008.0607.
26. S. Robin et al., "Genetic Diversity of Canine Olfactory Receptors," *BMC Genomics* 10, no. 21 (2009): 21.
27. A. Goldblatt, I. Gazit, and J. Terkel, "Olfaction and Explosives Detector Dogs," in *Canine Ergonomics: The Science of Working Dogs*, ed. W. S. Helton (CRC Press, 2009), 135–74.
28. S. Tacher et al., "Olfactory Receptor Sequence Polymorphism within and between Breeds of Dogs," *Journal of Heredity* 96 (2005): 812–16, doi:10.1093/jhered/esi113.
29. Quignon, Robin, and Galibert, "Canine Olfactory Genetics."
30. Syrotuck, *Scent and the Scenting Dog*; V. Fenton, "The Use of Dogs in Search, Rescue and Recovery," *Journal of Wilderness Medicine* 3 (1992): 292–300, doi:10.1580/0953-9859-3.3.292.
31. Syrotuck, *Scent and the Scenting Dog*.
32. D. B. Walker et al., "Naturalistic Quantification of Canine Olfactory Sensitivity," *Applied Animal Behaviour Science* 97 (2006): 241–54, doi:10.1016/j.applanim.2005.07.009; J. C. Walker et al., "Human Odor Detectability: New Methodology Used to Determine Threshold and Variation," *Chemical Senses* 28 (2003): 817–26, doi:10.1093/chemse/bjg075.
33. S. M. Stejskal, *Death, Decomposition, and Detector Dogs* (CRC Press, 2013).
34. Syrotuck, *Scent and the Scenting Dog*.
35. Goldblatt, Gazit, and Terkel, "Olfaction and Explosives Detector Dogs."
36. R. L. Atkinson et al., *Introduction to Psychology* (Harcourt Brace Jovanovich, 1990).
37. Quignon, Robin, and Galibert, "Canine Olfactory Genetics."
38. Quignon et al., "Genetics of Canine Olfaction and Receptor Diversity."
39. Syrotuck, *Scent and the Scenting Dog* (Barkleigh Productions, 1972); Fenton, "The Use of Dogs in Search, Rescue and Recovery"; M. Safran et al., "Human Gene-Centric Databases at the Weizmann Institute of Science: GeneCards, UDB, CroW 21 and HORDE," *Nucleic Acids Research* 31 (2003): 142–46, doi:10.1093/nar/gkg050; Syrotuck, *Scent and the Scenting Dog*.
40. M. D. Pearsall and M. D. Verbruggen, *Scent: Training to Track, Search and Rescue* (Alpine Publications, 1982).
41. Walker et al., "Naturalistic Quantification of Canine Olfactory Sensitivity"; Y. Nagata and N. Takeuchi, "Measurement of Odor Threshold by Triangle Odor Bag Method," *Bulletin of Japan Environmental Sanitation Center* 17 (1990): 77–89.
42. S. Bulanda with L. Bulanda, *Ready! Training the Search and Rescue Dog*, 2nd ed. (Kennel Club Books, 2010).
43. B. Adams et al., "Use of a Delayed Non-Matching to Position Task to Model Age-Dependent Cognitive Decline in the Dog," *Behavioural Brain Research* 108 (2000):

47–56, doi:10.1016/s0166-4328(99)00132-1; D. L. Wells and P. G. Hepper, "Directional Tracking in the Domestic Dog, *Canis familiaris*," *Applied Animal Behaviour Science* 84 (2003): 297–305, doi:10.1016/j.applanim.2003.08.009.

44. L. J. Wallis et al., "Lifespan Development of Attentiveness in Domestic Dogs: Drawing Parallels with Humans," *Frontiers in Psychology* 5 (2014), doi:10.3389/fpsyg.2014.00071.

45. D. G. Moulton, E. H. Ashton, and J. T. Eayrs, "Studies in Olfactory Acuity. 4. Relative Detectability of N-Alipatic Acids in Dog," *British Journal of Animal Behaviour* 8 (1960): 129–33.

46. D. Grandjean and F. Haymann, *Hunde-encyklopedi* (RK Grafisk, 2010); M. B. Parlee, "Menstrual Rhythms in Sensory Processes: A Review of Fluctuations in Vision, Olfaction, Audition, Taste and Touch," *Psychological Bulletin* 93 (1983): 539–48.

47. J. Riva et al., "The Effects of Drug Detection Training on Behavioral Reactivity and Blood Neurotransmitter Levels in Drug Detection Dogs: A Preliminary Study," *Journal of Veterinary Behavior: Clinical Applications and Research* 7 (2012): 11–20, doi:10.1016/j.jveb.2011.04.002.

48. L. J. Myers et al., "Dysfunction of Sense of Smell Caused by Canine Parainfluenza Virus-Infection in Dogs," *American Journal of Veterinary Research* 49 (1988): 188–90.

49. L. Gunnarsson et al., "Experimental Infection of Dogs with the Nasal Mite *Pneumonyssoides caninum*," *Veterinary Parasitology* 77 (1998): 179–86, doi:10.1016/s0304-4017(98)00104-6; W. P. Bredal, "The Prevalence of Nasal Mite (*Pneumonyssoides caninum*) Infection in Norwegian Dogs," *Veterinary Parasitology* 76 (1998): 233–37, doi:10.1016/s0304-4017(97)00204-5.

50. L. K. Gunnarsson et al., "Prevalence of *Pneumonyssoides caninum* Infection in Dogs in Sweden," *Journal of the American Animal Hospital Association* 37 (2001): 331–37; B. Gjerde, e-mail message to author, December 5, 2013.

51. A. King, "The Nose Knows," *New Scientist* 24 (August 2013): 40–43.

52. E. B. Arnett, "A Preliminary Evaluation on the Use of Dogs to Recover Bat Fatalities at Wind Energy Facilities," *Wildlife Society Bulletin* 34 (2006): 1440–45, doi:10.2193/0091-7648(2006)34[1440:apeotu]2.0.co;2; S. K. Wasser et al., "Scat Detection Dogs in Wildlife Research and Management: Application to Grizzly and Black Bears in the Yellowhead Ecosystem, Alberta, Canada," *Canadian Journal of Zoology/Revue canadienne de zoologie* 82 (2004): 475–92, doi:10.1139/z04-020; M. E. Cablk et al., "Olfaction-Based Detection Distance: A Quantitative Analysis of How Far Away Dogs Recognize Tortoise Odor and Follow It to Source," *Sensors* 8 (2008): 2208–22, doi:10.3390/s8042208; J. Paula et al., "Dogs as a Tool to Improve Bird-Strike Mortality Estimates at Wind Farms," *Journal for Nature Conservation* 19 (2011): 202–8, doi:10.1016/j.jnc.2011.01.002.

53. D. Komar, "The Use of Cadaver Dogs in Locating Scattered, Scavenged Human Remains: Preliminary Field Test Results," *Journal of Forensic Sciences* 44 (1999): 405–8.

54. Craven, Paterson, and Settles, "The Fluid Dynamics of Canine Olfaction."
55. G. S. Settles, D. A. Kester, and L. Dodson-Dreibelbis, "The External Aerodynamics of Sniffing," in *Sensors and Sensing in Biology and Engineering* (Springer, 2003).
56. J. B. Steen et al., "Olfaction in Bird Dogs during Hunting," *Acta Physiologica Scandinavica* 157 (1996): 115–19, doi:10.1046/j.1365-201X.1996.479227000.x.
57. A. Thesen, J. B. Steen, and K. B. Doving, "Behavior of Dogs during Olfactory Tracking," *Journal of Experimental Biology* 180 (1993): 247–51.
58. G. Settles and D. Kester, "Aerodynamic Sampling for Landmine Trace Detection," *SPIE Aerosense* 4294, paper 108 (2001).
59. I. Gazit and J. Terkel, "Domination of Olfaction over Vision in Explosives Detection by Dogs," *Applied Animal Behaviour Science* 82 (2003): 65–73, doi:10.1016/s0168-1591(03)00051-0.
60. King, "The Nose Knows"; E. K. Altom et al., "Effect of Dietary Fat Source and Exercise on Odorant-Detecting Ability of Canine Athletes," *Research in Veterinary Science* 75 (2003): 149–55, doi:10.1016/s0034-5288(03)00071-7.
61. I. Gazit and J. Terkel, "Explosives Detection by Sniffer Dogs Following Strenuous Physical Activity," *Applied Animal Behaviour Science* 81 (2003): 149–61, doi:10.1016/s0168-1591(02)00274-5.
62. D. A. Smith et al., "Detection and Accuracy Rates of Dogs Trained to Find Scats of San Joaquin Kit Foxes (*Vulpes macrotis mutica*)," *Animal Conservation* 6 (2003): 339–46, doi:10.1017/s136794300300341x.
63. A. E. Snovak, *Guide to Search and Rescue Dogs* (Barron's Educational Series, 2004).
64. Syrotuck, *Scent and the Scenting Dog.*
65. Fenton, "The Use of Dogs in Search, Rescue and Recovery."
66. J. E. King, R. F. Becker, and J. E. Markee, "Studies on Olfaction Discrimination in Dogs: (3) Ability to Detect Human Odour Trace," *Animal Behaviour* 12 (1964): 311–15.
67. Settles, Kester, and Dodson-Dreibelbis, "The External Aerodynamics of Sniffing."
68. K. Tonosaki and D. Tucker, "Responsiveness of the Olfactory Receptor-Cells in Dog to Some Odors," *Comparative Biochemistry and Physiology A-Physiology* 81 (1985): 7–13.
69. Lawson et al., "A Computational Study of Odorant Transport and Deposition in the Canine Nasal Cavity."
70. M. Siniscalchi et al., "Sniffing with the Right Nostril: Lateralization of Response to Odour Stimuli by Dogs," *Animal Behaviour* 82 (2011): 399–404, doi:10.1016/j.anbehav.2011.05.020.
71. Henshaw, *A Tour of the Senses.*
72. J. D. Mainland et al., "The Missense of Smell: Functional Variability in the Human Odorant Receptor Repertoire," *Nature Neuroscience* 17 (2014): 114–20, doi:10.1038/nn.3598 http://www.nature.com/neuro/journal/v17/n1/abs/nn.3598.html#supplementary-information.
73. Henshaw, *A Tour of the Senses.*

74. I. Croy, S. Nordin, and T. Hummel, "Olfactory Disorders and Quality of Life—an Updated Review," *Chemical Senses* 39 (2014): 185–94.
75. M. Whisman et al., "Odorant Evaluation: A Study of Ethanethiol and Tetrahdrothiophene as Warning Agents in Propane," *Environmental Science & Technology* 12 (1978): 1285–88.
76. L. Sela and N. Sobel, "Human Olfaction: A Constant State of Change-Blindness," *Experimental Brain Research* 205 (2010): 13–29, doi:10.1007/s00221-010-2348-6; C. Zelano and N. Sobel, "Humans as an Animal Model for Systems-Level Organization of Olfaction," *Neuron* 48 (2005): 431–54, doi:10.1016/j.neuron.2005.10.009.
77. C. Bushdid et al., "Humans Can Discriminate More than 1 Trillion Olfactory Stimuli," *Science* 343 (2014): 1370–72, doi:10.1126/science.1249168.
78. Sela, and Sobel, "Human Olfaction."
79. D. L. Wells and P. G. Hepper, "The Discrimination of Dog Odours by Humans," *Perception* 29 (2000): 111–15, doi:10.1068/p2938.
80. A. R. Radulescu and L. R. Mujica-Parodi, "Human Gender Differences in the Perception of Conspecific Alarm Chemosensory Cues," *PLOS ONE* 8 (2013), doi:10.1371/journal.pone.0068485.
81. S. B. Olsson, J. Barnard, and L. Turri, "Olfaction and Identification of Unrelated Individuals: Examination of the Mysteries of Human Odor Recognition," *Journal of Chemical Ecology* 32 (2006): 1635–45, doi:10.1007/s10886-006-9098-8.
82. M. Kaitz et al., "Mothers' Recognition of Their Newborns by Olfactory Cues," *Developmental Psychobiology* 20 (1987): 587–91, doi:10.1002/dev.420200604; R. H. Porter and J. Winberg, "Unique Salience of Maternal Breast Odors for Newborn Infants," *Neuroscience and Biobehavioral Reviews* 23 (1999): 439–49, doi:10.1016/s0149-7634(98)00044-x; S. Vaglio, in *Vitamins and Hormones: Pheromones*, vol. 83, *Vitamins and Hormones*, ed. G. Litwack (Elsevier Academic Press, 2010), 289–304.
83. R. H. Porter et al., "Recognition of Kin through Characteristic Body Odors," *Chemical Senses* 11 (1986): 389–95, doi:10.1093/chemse/11.3.389; B. Schaal and R. H. Porter, "Microsmatic Humans Revisited: The Generation and Perception of Chemical Signals," *Advances in the Study of Behavior* 20 (1991): 135–99; Porter et al., "Recognition of Kin through Characteristic Body Odors"; C. Ferdenzi, B. Schaal, and S. C. Roberts, "Family Scents: Developmental Changes in the Perception of Kin Body Odor?" *Journal of Chemical Ecology* 36 (2010): 847–54, doi:10.1007/s10886-010-9827-x.
84. C. Wedekind and S. Furi, "Body Odour Preferences in Men and Women: Do They Aim for Specific MHC Combinations or Simply Heterozygosity?" *Proceedings of the Royal Society B: Biological Sciences* 264 (1997): 1471–79; C. Wedekind et al., "MHC-Dependent Mate Preferences in Humans," *Proceedings of the Royal Society B-Biological Sciences* 260 (1995): 245–49, doi:10.1098/rspb.1995.0087.
85. Wedekind et al., "MHC-Dependent Mate Preferences in Humans"; S. C. Roberts et al., "MHC-Correlated Odour Preferences in Humans and the Use of Oral Contraceptives," *Proceedings of the Royal Society B: Biological Sciences* 275 (2008): 2715–22, doi:10.1098/rspb.2008.0825.

86. J. Havlíček et al., "Non-Advertised Does Not Mean Concealed: Body Odour Changes across the Human Menstrual Cycle," *Ethology* 112 (2006): 81–90, doi:10.1111/j.1439-0310.2006.01125.x.
87. A. Arzi et al., "Humans Can Learn New Information during Sleep," *Nature Neuroscience* 15 (2012): 1460–65, doi:10.1038/nn.3193.
88. Sela and Sobel, "Human Olfaction"; Zelano and Sobel, "Humans as an Animal Model for Systems-Level Organization of Olfaction."
89. G. M. Shepherd, "The Human Sense of Smell: Are We Better than We Think?" *PLoS Biology* 2 (2004): 572–75, doi:10.1371/journal.pbio.0020146.
90. D. R. Adams and M. D. Wiekamp, "The Canine Vomeronasal Organ," *Journal of Anatomy* 138 (1984): 771–87; I. Salazar et al., "Structural, Morphometric, and Immunohistological Study of the Accessory Olfactory-Bulb in the Dog," *Anatomical Record* 240 (1994): 277–85, doi:10.1002/ar.1092400216; I. Salazar, P. C. Barber, and J. M. Cifuentes, "Anatomical and Immunohistological Demonstration of the Primary Neural Connections of the Vomeronasal Organ in the Dog," *Anatomical Record* 233 (1992): 309–13, doi:10.1002/ar.1092330214.
91. L. Jacobson, "Anatomisk beskrivelse over et nyt organ i husdyrenes næse," *Veterinær Selskapets Skrifter* 2 (1813): 209–46.
92. G. A. A. Schoon and R. Haak, *K9 Suspect Discrimination* (Detseling Enterprises Ltd., 2002).
93. Evans and Lahunta, *Miller's Anatomy of the Dog*, 4th ed.
94. D. Berthoud et al., "The Effect of Castrating Male Dogs on Their Use of the Vomeronasal Organ When Investigating Conspecific Urine Deposits," Third Canine Science Forum (Barcelona, July 25–27, 2012).
95. J. Bradshaw, *In Defence of Dogs* (Penguin, 2011).
96. M. Goodwin, K. M. Gooding, and F. Regnier, "Sex-Pheromone in the Dog," *Science* 203 (1979): 559–61, doi:10.1126/science.569903.
97. Fyrand, *Det gåtefulle språket*.
98. Berthoud et al., "The Effect of Castrating Male Dogs on Their Use of the Vomeronasal Organ."
99. G. M. Zucco et al., "Odor-Evoked Autobiographical Memories: Age and Gender Differences Along the Life Span," *Chemical Senses* 37 (2012): 179–89, doi:10.1093/chemse/bjr089.
100. M. Williams and J. M. Johnston, "Training and Maintaining the Performance of Dogs (*Canis familiaris*) on an Increasing Number of Odor Discriminations in a Controlled Setting," *Applied Animal Behaviour Science* 78 (2002): 55–65, doi:10.1016/s0168-1591(02)00081-3.
101. Schoon and Haak, *K9 Suspect Discrimination*.

CHAPTER THREE

1. S. Bulanda, e-mail message to author, March 18, 2013.
2. D. Grandjean and F. Haymann, *Hunde-encyklopedi* (RK Grafisk, 2010).

3. Ibid.
4. J. Smith, "Africa's New Poaching Police," *Men's Journal* (2012).
5. K. Most and G. H. Brückner, "Über voraussetzungen und den derzeitigen stand der nasenleistungen von hunden," *Zeitschrift für Hundeforschunge* 12 (1936): 9–30.
6. J. Schettler, *Red Dog Rising* (Alpine Publications, 2009).
7. W. G. Syrotuck, *Scent and the Scenting Dog* (Arner Publishing, 2000).
8. J. Schettler, *K-9 Trailing: The Straightest Path* (Alpine Publications, 2011).
9. Syrotuck, *Scent and the Scenting Dog*.
10. B. A. Edge, E. G. Paterson, and G. S. Settles, "Computational Study of the Wake and Contaminant Transport of a Walking Human," *Journal of Fluids Engineering, Transactions of the ASME* 127 (2005): 967–77, doi:10.1115/1.2013291.
11. J. I. Choi and J. R. Edwards, "Large Eddy Simulation and Zonal Modeling of Human-Induced Contaminant Transport," *Indoor Air* 18 (2008): 233–49, doi:10.1111/j.1600-0668.2008.00527.x.
12. Syrotuck, *Scent and the Scenting Dog*; K. Findley et al., "Topographic Diversity of Fungal and Bacterial Communities in Human Skin," *Nature* 498 (2013): 367–70, doi:10.1038/nature12171 http://www.nature.com/nature/journal/vaop/ncurrent/abs/nature12171.html#supplementary-information; H. C. Korting, A. Lukacs, and O. Braun-Falco, "Mikrobielle flora und geruch der gesunden mensclichen haut," *Hautarzt* 39 (1988): 564–68.
13. L. Dormont, J. M. Bessiere, and A. Cohuet, "Human Skin Volatiles: A Review," *Journal of Chemical Ecology* 39 (2013): 569–78, doi:10.1007/s10886-013-0286-z.
14. J. Fialová, S. C. Roberts, and J. Havlíček, "Is the Perception of Dietary Odour Cues Linked to Sexual Selection in Humans?" in *Chemical Signals in Vertebrates 12*, ed. M. East and M. Dehnhard (Springer Science+Business Media, 2013); J. Havlíček and P. Lenochova, "The Effect of Meat Consumption on Body Odor Attractiveness," *Chemical Senses* 31 (2006): 747–52, doi:10.1093/chemse/bjl017.
15. Dormont, Bessiere, and Cohuet, "Human Skin Volatiles"; S. Kippenberger et al.,"'Nosing Around' the Human Skin: What Information Is Concealed in Skin Odour?" *Experimental Dermatology* 21 (2012): 655–59, doi:10.1111/j.1600--0625.2012.01545.x.
16. S. J. Song et al., "Cohabiting Family Members Share Microbiota with One Another and with Their Dogs," *eLife* 2 (2013), e00458-e00458, doi:10.7554/eLife.00458.
17. J. Stokholm et al., "Living with Cat and Dog Increases Vaginal Colonization with *E. coli* in Pregnant Women," *PLOS ONE* 7 (2012), doi:10.1371/journal.pone.0046226.
18. R. R. Dunn et al., "Home Life: Factors Structuring the Bacterial Diversity Found within and between Homes," *PLOS ONE* 8 (2013), doi:10.1371/journal.pone.0064133.
19. M. Gallagher et al., "Analyses of Volatile Organic Compounds from Human Skin," *British Journal of Dermatology* 159 (2008): 780–91, doi:10.1111/j.1365--2133.2008.08748.x.

20. Dormont, Bessiere, and Cohuet, "Human Skin Volatiles."
21. Ibid.; Kippenberger et al., "'Nosing Around' the Human Skin"; A. Girod, R. Ramotowski, and C. Weyermann, "Composition of Fingermark Residue: A Qualitative and Quantitative Review," *Forensic Science International* 223 (2012): 10–24, doi:10.1016/j.forsciint.2012.05.018.
22. N. Goetz et al., "Detection and Identification of Volatile Compounds Evolved from Human-Hair and Scalp Using Headspace Gas-Chromatography," *Journal of the Society of Cosmetic Chemists* 39 (1988): 1–13.
23. Girod, Ramotowski, and Weyermann, "Composition of Fingermark Residue."
24. M. Kusano, E. Mendez, and K. G. Furton, "Comparison of the Volatile Organic Compounds from Different Biological Specimens for Profiling Potential," *Journal of Forensic Sciences* 58 (2013): 29–39, doi:10.1111/j.1556-4029.2012.02215.x.
25. J. C. Filiatre et al., "An Experimental Analysis of Olfactory Cues in Child-Dog Interaction," *Chemical Senses* 15 (1990): 679–89, doi:10.1093/chemse/15.6.679.
26. J. C. Filiatre, J. L. Millot, and A. Eckerlin, "Behavioral Variability of Olfactory Exploration of the Pet Dog in Relation to Human Adults," *Applied Animal Behaviour Science* 30 (1991): 341–50, doi:10.1016/0168-1591(91)90139-0.
27. Kippenberger et al., "'Nosing Around' the Human Skin."
28. U. Ellervik, *Ond kemi. Berättelser om människor, mord och molekyler* (Fri tanke forlag, 2011).
29. Kusano, Mendez, and Furton, "Comparison of the Volatile Organic Compounds from Different Biological Specimens for Profiling Potential."
30. Gallagher et al., "Analyses of Volatile Organic Compounds from Human Skin."
31. G. A. A. Schoon and R. Haak, *K9 Suspect Discrimination* (Detseling Enterprises Ltd., 2002).
32. Gallagher et al., "Analyses of Volatile Organic Compounds from Human Skin"; L. Dormont et al., "New Methods for Field Collection of Human Skin Volatiles and Perspectives for Their Application in the Chemical Ecology of Human-Pathogen-Vector Interactions," *Journal of Experimental Biology* 216 (2013): 2783–88, doi:10.1242/jeb.085936.
33. K. Ara et al., "Foot Odor Due to Microbial Metabolism and Its Control," *Canadian Journal of Microbiology* 52 (2006): 357–64, doi:10.1139/w05-130.
34. L. Fält et al., *Spårhunden och lukterna* (SWDI forlag, 2011); in English: *Tracking Dogs: Scents and Skills*, https://webshop.spesialsok.no/home/10-tracking-dogs-scent-and-skills.html.
35. A. M. Curran, P. A. Prada, and K. G. Furton, "The Differentiation of the Volatile Organic Signatures of Individuals through SPME-GC/MS of Characteristic Human Scent Compounds," *Journal of Forensic Sciences* 55 (2010): 50–57, doi:10.1111/j.1556-4029.2009.01236.x.
36. A. Girod and C. Weyermann, "Lipid Composition of Fingermark Residue and Donor Classification Using GC/MS," *Forensic Science International* 238C (2014): 68–82, doi:10.1016/j.forsciint.2014.02.020.

37. L. Löhner, "Über menschliche individual- und regionalgerüche," *Arch. Ges. Physiol.* 202 (1924): 25–45.
38. J. E. King et al., "Studies on Olfaction Discrimination in Dogs: (3) Ability to Detect Human Odour Trace," *Animal Behaviour* 12 (1964): 311–15.
39. Santariova, "Ability of Scent Identification Dogs to Detect Individual Human Odors," Third Canine Science Forum (Barcelona, July 25–27, 2012).
40. P. Vyplelová et al., "Individual Human Odor Fallout as Detected by Trained Canines," *Forensic Science International* 234 (2014): 13–15, doi:http://dx.doi.org/10.1016/j.forsciint.2013.10.018.
41. T. Jezierski et al., "Do Trained Dogs Discriminate Individual Body Odors of Women Better than Those of Men?" *Journal of Forensic Sciences* 57 (2012): 647–53, doi:10.1111/j.1556-4029.2011.02029.x.
42. A. M. Curran et al., "The Frequency of Occurrence and Discriminatory Power of Compounds Found in Human Scent across a Population Determined by SPME-GCMS," *Journal of Chromatography B: Analytical Technologies in the Biomedical and Life Sciences* 846 (2007): 86–97, doi:10.1016/j.jchromb.2006.08.039.
43. J. Kloek, "The Smell of Some Steroid Sex Hormones and Their Metabolites: Reflections and Experiments Concerning the Significance of Smell for the Mutual Relations of the Sexes," *Psychiatria, Neurologia, Neurochirurgia* 64 (1961): 309–44.
44. A. M. Curran, P. A. Prada, and K. G. Furton, "Canine Human Scent Identifications with Post-Blast Debris Collected from Improvised Explosive Devices," *Forensic Science International* 199 (2010): 103–8, doi:10.1016/j.forsciint.2010.03.021.
45. J. B. Steen and E. Wilsson, "How Do Dogs Determine the Direction of Tracks?," *Acta Physiologica Scandinavica* 139 (1990): 531–34, doi:10.1111/j.1748-1716.1990.tb08956.x.
46. P. G. Hepper and D. L. Wells, "How Many Footsteps Do Dogs Need to Determine the Direction of an Odour Trail?" *Chemical Senses* 30 (2005): 291–98, doi:10.1093/chemse/bji023.
47. Ibid.
48. A. Thesen, J. B. Steen, and K. B. Doving, "Behavior of Dogs during Olfactory Tracking," *Journal of Experimental Biology* 180 (1993): 247–51; Hepper and Wells, "How Many Footsteps Do Dogs Need to Determine the Direction of an Odour Trail?"
49. Steen and Wilsson, "How Do Dogs Determine the Direction of Tracks?"
50. Hepper and Wells, "How Many Footsteps Do Dogs Need to Determine the Direction of an Odour Trail?"
51. D. L. Wells and P. G. Hepper, "Directional Tracking in the Domestic Dog, *Canis familiaris*," *Applied Animal Behaviour Science* 84 (2003): 297–305, doi:10.1016/j.applanim.2003.08.009; Steen and Wilsson, "How Do Dogs Determine the Direction of Tracks?"
52. Wells and Hepper, "Directional Tracking in the Domestic Dog, *Canis familiaris*."
53. D. Müller-Schwarze, *Chemical Ecology of Vertebrates* (Cambridge University Press, 2006); Lisberg and Snowdon, "Effects of Sex, Social Status and Gonadectomy on Countermarking by Domestic Dogs, *Canis familiaris*."

54. C. Murphy, "Age-Associated Differences in Memories for Odors," in *Memory for Odors*, ed. F. R. Schab and R. G. Crowder (Lawrence Erlbaum Associates, 1995), 109–31; C. Sinding, L. Puschmann, and T. Hummel, "Is the Age-Related Loss in Olfactory Sensitivity Similar for Light and Heavy Molecules?" *Chemical Senses* 39 (2014): 383–90, doi:10.1093/chemse/bju004.
55. H. E. Salvin et al., "Development of a Novel Paradigm for the Measurement of Olfactory Discrimination in Dogs (*Canis familiaris*): A Pilot study," *Journal of Veterinary Behavior: Clinical Applications and Research* 7 (2012): 3–10, doi:10.1016/j.jveb.2011.04.005.
56. T. Hirai et al., "Age-Related Changes in the Olfactory System of Dogs," *Neuropathology and Applied Neurobiology* 22 (1996): 531–39.
57. Wells and Hepper, "Directional Tracking in the Domestic Dog, *Canis familiaris*."
58. D. L. Wells, e-mail message to author, April 10, 2014.
59. J. Porter et al., "Mechanisms of Scent-Tracking in Humans," *Nature Neuroscience* 10 (2007): 27–29, doi:10.1038/nn1819.
60. Ibid.
61. J. M. Gardiner and J. Atema, "The Function of Bilateral Odor Arrival Time Differences in Olfactory Orientation of Sharks," *Current Biology* 20 (2010): 1187–91, doi:10.1016/j.cub.2010.04.053.
62. B. A. Craven, E. G. Paterson, and G. S. Settles, "The Fluid Dynamics of Canine Olfaction: Unique Nasal Airflow Patterns as an Explanation of Macrosmia," *Journal of the Royal Society Interface* 7 (2010): 933–43, doi:10.1098/rsif.2009.0490.
63. G. Romanes, "Experiments on the Sense of Smell in Dogs," *Nature* 36 (1887): 273–74.
64. E. A. Archie and K. R. Theis, "Animal Behaviour Meets Microbial Ecology," *Animal Behaviour* 82 (2011): 425–36, doi:10.1016/j.anbehav.2011.05.029; P. A. Prada and K. G. Furton, "Human Scent Detection: A Review of Its Developments and Forensic Applications," *Revista de Ciencias Forenses* 1 (2008): 81–87.
65. Curran, Prada and Furton, "The Differentiation of the Volatile Organic Signatures of Individuals through SPME-GC/MS of Characteristic Human Scent Compounds"; M. D. Thom, R. J. Beynon and J. L. Hurst, "The Role of the Major Histocompatibility Complex in Scent Communication," in *Chemical Signals in Vertebrates* 10, ed. R. T. Mason, M. P. Le Master, and D. Müller-Schwarze (Springer, 2005), 173–82.
66. Kusano, Mendez, and Furton, "Comparison of the Volatile Organic Compounds from Different Biological Specimens for Profiling Potential."
67. R. Thornhill et al., "Major Histocompatibility Complex Genes, Symmetry, and Body Scent Attractiveness in Men and Women," *Behavioral Ecology* 14 (2003): 668–78, doi:10.1093/beheco/arg043; P. S. C. Santos et al., "New Evidence that the MHC Influences Odor Perception in Humans: A Study with 58 Southern Brazilian Students," *Hormones and Behavior* 47 (2005): 384–88, doi:10.1016/j.yhbeh.2004.11.005.

68. S. Bjering, I. I. Deinboll, and J. Mæhlen, "Luktesansen," *Tidsskrift Norske Lægeforening* 30 (2000): 3719–25.
69. T. Boehm and F. Zufall, "MHC Peptides and the Sensory Evaluation of Genotype," *Trends in Neurosciences* 29 (2006): 100–107, doi:10.1016/j.tins.2005.11.006.
70. Kippenberger et al., "'Nosing Around' the Human Skin."
71. M. Troccaz et al., "Gender-Specific Differences between the Concentrations of Nonvolatile (R)/(S)-3-Methyl-3-Sulfanylhexan-1-Ol and (R)/(S)-3-Hydroxy-3-Methyl-Hexanoic Acid Odor Precursors in Axillary Secretions," *Chemical Senses* 34 (2009): 203–10, doi:10.1093/chemse/bjn076.
72. D. J. Penn et al., "Individual and Gender Fingerprints in Human Body Odour," *Journal of the Royal Society Interface* 4 (2007): 331–40, doi:10.1098/rsif.2006.0182.
73. Syrotuck, *Scent and the Scenting Dog.*
74. Kippenberger et al. "'Nosing Around' the Human Skin."
75. Kippenberger et al., "'Nosing Around' the Human Skin"; C. M. Willis et al., "Olfactory Detection of Human Bladder Cancer by Dogs: Proof of Principle Study," *British Medical Journal* 329 (2004): 712–14A, doi:10.1136/bmj.329.7468.712.
76. Schettler, *Red Dog Rising*; Schettler, *K-9 Trailing.*
77. S. Gelstein et al., "Human Tears Contain a Chemosignal," *Science* 331 (2011): 226–30, doi:10.1126/science.1198331.
78. Schettler, *Red Dog Rising*; Schettler, *K-9 Trailing.*
79. Curran, Prada, and Furton, "The Differentiation of the Volatile Organic Signatures of Individuals through SPME-GC/MS of Characteristic Human Scent Compounds."
80. Kippenberger et al., "'Nosing Around' the Human Skin"; Troccaz et al., "Gender-Specific Differences between the Concentrations of Nonvolatile (R)/(S)-3-Methyl-3-Sulfanylhexan-1-Ol and (R)/(S)-3-Hydroxy-3-Methyl-Hexanoic Acid Odor Precursors in Axillary Secretions."
81. Syrotuck, *Scent and the Scenting Dog*; A. Martin et al., "A Functional ABCC11 Allele Is Essential in the Biochemical Formation of Human Axillary Odor," *Journal of Investigative Dermatology* 130 (2010): 529–40, doi:10.1038/jid.2009.254.
82. Kippenberger et al., "'Nosing Around' the Human Skin"; M. Gallagher et al., "Analyses of Volatile Organic Compounds from Human Skin"; S. Mitro et al., "The Smell of Age: Perception and Discrimination of Body Odors of Different Ages," *PLOS ONE* 7 (2012), doi:10.1371/journal.pone.0038110.
83. Archie and Theis, "Animal Behaviour Meets Microbial Ecology"; G. E. Weisfeld et al., "Possible Olfaction-Based Mechanisms in Human Kin Recognition and Inbreeding Avoidance," *Journal of Experimental Child Psychology* 85 (2003): 279–95, doi:10.1016/s0022-0965(03)00061-4.
84. B. Schaal and R. H. Porter, "Microsmatic Humans Revisited: The Generation and Perception of Chemical Signals," *Advances in the Study of Behavior* 20 (1991): 135–99; D. Singh and P. M. Bronstad, "Female Body Odour Is a Potential Cue to Ovulation," *Proceedings of the Royal Society B: Biological Sciences* 268 (2001): 797–801.

85. Kippenberger et al., "'Nosing Around' the Human Skin"; M. Shirasu and K. Touhara, "The Scent of Disease: Volatile Organic Compounds of the Human Body Related to Disease and Disorder," *Journal of Biochemistry* 150 (2011): 257–66, doi:10.1093/jb/mvr090; F. Prugnolle et al., "Infection and Body Odours: Evolutionary and Medical Perspectives," *Infection Genetics and Evolution* 9 (2009): 1006–9, doi:10.1016/j.meegid.2009.04.018.
86. D. Chen and J. Haviland-Jones, "Human Olfactory Communication of Emotion," *Perceptual and Motor Skills* 91 (2000): 771–81, doi:10.2466/pms.91.7.771-781; K. Ackerl, M. Atzmueller, and K. Grammer, "The Scent of Fear," *Neuroendocrinology Letters* 23 (2002): 79–84.
87. Schaal and Porter, "Microsmatic Humans Revisited."
88. G. S. Berns, A. M. Brooks, and M. Spivak, "Functional MRI in Awake Unrestrained Dogs," *PLOS ONE* 7 (2012), doi:10.1371/journal.pone.0038027.
89. G. Berns, e-mail message to author, November 21, 2013; G. S. Berns, A. Brooks, and M. Spivak, "Replicability and Heterogeneity of Awake Unrestrained Canine fMRI Responses," *PLOS ONE* 8 (2013), doi:10.1371/journal.pone.0081698.
90. G. Berns, *How Dogs Love Us: A Neuroscientist and His Adopted Dog Decode the Canine Brain* (Lake Union Publishing, 2013).
91. Berns, Brooks, and Spivak, "Functional MRI in Awake Unrestrained Dogs"; H. Jia et al., "Functional MRI of the Olfactory System in Conscious Dogs," *PLOS ONE* 9 (2014), doi:10.1371/journal.pone.0086362.
92. H. Kalmus, "The Discrimination by the Nose of the Dog of Individual Human Odours and in Particular of the Odours of Twins," *British Journal of Animal Behaviour* 3 (1955): 25–31, doi: 10.1068/p170549; P. G. Hepper, "The Discrimination of Human Odor by the Dog," *Perception* 17 (1988): 549–54, doi:10.1068/p170549; B. A. Sommerville et al., "The Use of Trained Dogs to Discriminate Human Scent," *Animal Behaviour* 46 (1993): 189–90, 10.1006/anbe.1993.1174; L. M. Harvey et al., "The Use of Bloodhounds in Determining the Impact of Genetics and the Environment on the Expression of Human Odortype," *Journal of Forensic Sciences* 51 (2006): 1109–14, doi:10.1111/j.1556-4029.2006.00231.x.
93. L. Pinc et al., "Dogs Discriminate Identical Twins," *PLOS ONE* 6 (2011), doi:e2070410.1371/journal.pone.0020704.
94. Ibid.
95. T. Thomassen, e-mail message to author, September 2013; http://www.id-hund.no/.

CHAPTER FOUR

1. M. Weisbord and K. Kachanoff, *Dogs with Jobs* (Pocket Books, 2000); K. Albrecht, *Dog Detectives: Train Your Dog to Find Lost Pets* (Dogwise Publishing, 2008).
2. S. D. Gosling, C. J. Sandy, and J. Pottert, "Personalities of Self-Identified 'Dog People' and 'Cat People,'" *Anthrozoos* 23 (2010): 213–22, doi:10.2752/175303710X12750451258850.

3. T. F. Garrity and L. Stallones, "Effects of Pet Contact in Human Well-Being," in *Companion Animals in Human Health*, ed. C. Wilson and D. C. Turner (Sage, 1998); D. A. Marcus, "The Science Behind Animal-Assisted Therapy," *Current Pain and Headache Reports* 17 (2013), doi:10.1007/s11916-013-0322-2.
4. K. Aspaas, "Vi som elsker hund," *Aftenposten* 3 (June 2012).
5. T. Rehn and L. J. Keeling, "The Effect of Time Left Alone at Home on Dog Welfare," *Applied Animal Behaviour Science* 129 (2011): 129–35, doi:10.1016/j.applanim.2010.11.015.
6. T. Rehn et al., "I Like My Dog, Does My Dog Like Me?" *Applied Animal Behaviour Science* 150 (2014): 65–73, doi:http://dx.doi.org/10.1016/j.applanim.2013.10.008.
7. Weisbord and Kachanoff, *Dogs with Jobs*.
8. Kat Albrecht, http://www.katalbrecht.com/.
9. M. M. Bouzga and A. G. Skalleberg, "Hundeprosjektet," *Bevis* 3 (2013): 23–25.
10. V. Hart et al., "Dogs Are Sensitive to Small Variations of the Earth's Magnetic Field," *Frontiers in Zoology* 10 (2013): 80.
11. Check the following books for more information about scent organs and scent marking behavior: D. Müller-Schwarze, *Chemical Ecology of Vertebrates* (Cambridge University Press, 2006); T. D. Wyatt, *Pheromones and Animal Behavior: Chemical Signals and Signatures* (Cambridge University Press, 2014).
12. Müller-Schwarze, *Chemical Ecology of Vertebrates*; M. R. Conover, *Predator-Prey Dynamics: The Role of Olfaction* (CRC Press, 2007).
13. W. G. Syrotuck, *Scent and the Scenting Dog* (Arner Publishing, 2000).
14. L. Fält et al., *Spårhunden och lukterna* (SWDI forlag, 2011); in English: *Tracking Dogs: Scents and Skills*, https://webshop.spesialsok.no/home/10-tracking-dogs-scent-and-skills.html.
15. F. W. Christiansen, *Hundes og ulved adfærd* (Forlaget Tro-fast ApS, 2007).
16. B. Hare, e-mail message to author, May 28, 2014; M. Bekoff, e-mail message to author, June 2, 2014.
17. H. E. Evans and G. C. Christensen, "The Urogenital System," in *Miller's Anatomy of the Dog*, 3rd ed. (W. B. Saunders, 1993), 494–558.
18. T. Aoki and M. Wada, "Functional Activity of the Sweat Glands in the Hairy Skin of the Dog," *Science* 114 (1951): 123–24.
19. S. W. Nielsen, "Glands of the Canine Skin: Morphology and Distribution," *American Journal of Veterinary Research* 14 (July 1953): 448–54; J. G. Speed, "Sweat-Glands of the Dog," *Veterinary Journal* 97 (1941): 252–56.
20. Evans and Christensen, "The Urogenital System," in *Miller's Anatomy of the Dog*, 3rd ed.
21. Nielsen, "Glands of the Canine Skin"; J. E. Lovell and R. Getty, "The Hair Follicle, Epidermis, Dermis, and Skin Glands of the Dog," *American Journal of Veterinary Research* (October 1957): 873–85.
22. E. S. Albone, *Mammalian Semiochemistry: The Investigation of Chemical Signals between Mammals* (John Wiley & Sons, 1984).

23. M. Bekoff, "Ground Scratching by Male Domestic Dogs: A Composite Signal," *Journal of Mammalogy* 60 (1979): 847–48.
24. P. G. Hepper, "Long-Term Retention of Kinship Recognition Established during Infancy in the Domestic Dog," *Behavioural Processes* 33 (1994): 3–14, doi:10.1016/0376-6357(94)90056-6.
25. A. E. Lisberg and C. T. Snowdon, "Effects of Sex, Social Status and Gonadectomy on Countermarking by Domestic Dogs, *Canis familiaris*," *Animal Behaviour* 81 (2011): 757–64, doi:10.1016/j.anbehav.2011.01.006; I. F. Dunbar, "Olfactory Preferences in Dogs: The Response of Male and Female Beagles to Conspecific Urine," *Behavioral Biology* 20 (1978): 471–81, doi:10.1016/s0091-6773(77)91079-3.
26. M. Bekoff, "Scent-Marking by Free-Ranging Domestic Dogs: Olfactory and Visual Components," *Biology of Behaviour* 4 (1979): 123–39.
27. D. Grandjean and F. Haymann, *Hunde-encyklopedi* (RK Grafisk, 2010).
28. F. A. Beach, "Effects of Gonadal Hormones on Urinary Behavior in Dogs," *Physiology & Behavior* 12 (1974): 1005–13, doi:10.1016/0031-9384(74)90148-6; E. Ranson and F. A. Beach, "Effects of Testosterone on Ontogeny of Urinary Behavior in Male and Female Dogs," *Hormones and Behavior* 19 (1985): 36–51, doi:10.1016/0018-506x(85)90004-2.
29. R. H. Sprague and J. J. Anisko, "Elimination Patterns in the Laboratory Beagle," *Behaviour* 47 (1973): 257–67.
30. A. E. Lisberg and C. T. Snowdon, "The Effects of Sex, Gonadectomy and Status on Investigation Patterns of Unfamiliar Conspecific Urine in Domestic Dogs, *Canis familiaris*," *Animal Behaviour* 77 (2009): 1147–54, doi:10.1016/j.anbehav.2008.12.033.
31. Lisberg and Snowdon, "Effects of Sex, Social Status and Gonadectomy on Countermarking by Domestic Dogs, *Canis familiaris*"; I. Dunbar and M. Buehler, "A Masking Effect of Urine from Male Dogs," *Applied Animal Ethology* 6 (1980): 297–301.
32. Ibid.
33. A. L. Kvam, *Nesearbeid for hund. Skritt for skritt fra godbitsøk til sporsøk* (Tun forlag AS, 2005).
34. F. Rabelais, *Pantagruel-de torstigas konung* (Fachlcrantz & Gumaelius, 1945).
35. R. Campbell-Palmer and F. Rosell, "Conservation of the Eurasian Beaver *Castor fiber*: An Olfactory Perspective," *Mammal Review* 40 (2010): 293–312, doi:10.1111/j.1365-2907.2010.00165.x.
36. M. Bekoff, "Observations of Scent-Marking and Discriminating Self from Others by a Domestic Dog (*Canis familiaris*): Tales of Displaced Yellow Snow," *Behavioural Processes* 55 (2001): 75–79, doi:10.1016/s0376-6357(01)00142-5.
37. K. Taylor and D. S. Mills, "A Placebo-Controlled Study to Investigate the Effect of Dog Appeasing Pheromone and Other Environmental and Management Factors on the Reports of Disturbance and House Soiling during the Night in Recently Adopted Puppies (*Canis familiaris*)," *Applied Animal Behaviour Science* 105 (2007): 358–68.

38. O. Elliot and J. P. Scott, "The Development of Emotional Distress Reactions to Separation in Puppies," *Journal of Genetic Psychology* 99 (1961): 3–22.
39. E. Gaultier et al., "Efficacy of Dog-Appeasing Pheromone in Reducing Stress Associated with Social Isolation in Newly Adopted Puppies," *Veterinary Record* 163 (2008): 73–80.
40. G. Sheppard and D. S. Mills, "Evaluation of Dog-Appeasing Pheromone as a Potential Treatment for Dogs Fearful of Fireworks," *Veterinary Record* 152 (2003): 432–36; E. Gaultier et al., "Comparison of the Efficacy of a Synthetic Dog-Appeasing Pheromone with Clomipramine for the Treatment of Separation-Related Disorders in Dogs," *Veterinary Record* 156 (2005): 533–38; Y. M. Kim et al., "Efficacy of Dog-Appeasing Pheromone (DAP) for Ameliorating Separation-Related Behavioral Signs in Hospitalized Dogs," *Canadian Veterinary Journal/Revue vétérinaire canadienne* 51 (2010): 380–84; M. G. Estelles and D. S. Mills, "Signs of Travel-Related Problems in Dogs and Their Response to Treatment with Dog-Appeasing Pheromone," *Veterinary Record* 159 (2006): 143–48; S. Denenberg and G. M. Landsberg, "Effects of Dog-Appeasing Pheromones on Anxiety and Fear in Puppies during Training and on Long-Term Socialization," *Journal of the American Veterinary Medical Association* 233 (2008): 1874–82; E. Tod, D. Brander, and N. Waran, "Efficacy of Dog Appeasing Pheromone in Reducing Stress and Fear Related Behaviour in Shelter Dogs," *Applied Animal Behaviour Science* 93 (2005): 295–308, doi:10.1016/j.applamin.2005.01.007.
41. Taylor and Mills, "A Placebo-Controlled Study to Investigate the Effect of Dog Appeasing Pheromone."
42. L. Graham, D. L. Wells, and P. G. Hepper, "The Influence of Olfactory Stimulation on the Behaviour of Dogs Housed in a Rescue Shelter," *Applied Animal Behaviour Science* 91 (2005): 143–53, doi:10.1016/j.applanim.2004.08.024.
43. A. Horowitz, J. Hecht, and A. Dedriek, "Smelling More or Less: Investigating the Olfactory Experience of the Domestic Dog," *Learning and Motivation* 44 (2013): 207–17.
44. Evans and Christensen, "The Urogenital System," in *Miller's Anatomy of the Dog*, 3rd ed.
45. W. J. Bacha and L. M. Bacha, *Color Atlas of Veterinary Histology*, 2nd ed. (Lippincott Williams & Wilkins, 2000).
46. R. L. Doty and I. Dunbar, "Color, Odor, Consistency and Secretion Rate of Anal Sac Secretion from Male, Female and Early Androgenizied Female Beagles. *American Journal of Veterinary Research* 35 (1974): 729–31.
47. W. Montagna and H. F. Parks, "A Histochemical Study of the Glands of the Anal Sac of the Dog," *Anatomical Record* 100 (1948): 297–317.
48. A. M. Lake et al., "Gross and Cytological Characteristics of Normal Canine Anal-Sac Secretions," *Journal of Veterinary Medicine Series A: Physiology Pathology Clinical Medicine* 51 (2004): 249–53, doi:10.1111/j.1439-0442.2004.00629.x.
49. C. S. Asa et al., "Deposition of Anal-Sac Secretions by Captive Wolves (*Canis lupus*)," *Journal of Mammalogy* 66 (1985): 89–93, doi:10.2307/1380960.

50. Müller-Schwarze, *Chemical Ecology of Vertebrates.*
51. S. Natynczuk, J. W. S. Bradshaw, and D. W. Mcdonald, "Chemical Constituents of the Anal Sacs of Domestic Dogs," *Biochemical Systematics and Ecology* 17 (1989): 83–87, doi:10.1016/0305-1978(89)90047-1.
52. C. S. Hedlund and T. W. Fossum, "Surgery of the Perineum, Rectum, and Anus," in *Small Animal Surgery*, ed. T. W. Fossum, 3rd ed. (Mosby/Elsevier, 2007), 498–530.
53. D. J. James et al., "Comparison of Anal Sac Cytological Findings and Behaviour in Clinically Normal Dogs and Those Affected with Anal Sac Disease," *Veterinary Dermatology* 22 (2011): 80–87, doi:10.1111/j.1365-3164.2010.00916.x.
54. Ibid.
55. S. A. Shabadash and T. I. Zelikina, "The Tail Gland of Canids," *Biology Bulletin* 31 (2004): 367–76, doi:10.1023/B:BIBU.0000036941.18383.bd; D. W. Scott and T. J. Reimers, "Tail Gland and Perianal Gland Hyperplasia Associated with Testicular Neoplasia and Hypertestosteronemia in a Dog," *Canine Practice* 13 (1986): 15–17.
56. F. Al-Bagdadi, "The Integument," in *Miller's Anatomy of the Dog*, ed. H. E. Evans and A. de Lahunta, 4th ed. (Elsevier Saunders, 2013).
57. Grandjean and Haymann, *Hunde-encyklopedi.*
58. Sprague and Anisko, "Elimination Patterns in the Laboratory Beagle."
59. Dunbar, "Olfactory Preferences in Dogs."
60. Ranson and Beach, "Effects of Testosterone on Ontogeny of Urinary Behavior in Male and Female Dogs."
61. B. Pardo-Carmona et al., "Saliva Crystallisation as a Means of Determining Optimal Mating Time in Bitches," *Journal of Small Animal Practice* 51 (2010): 437–42, doi:10.1111/j.1748-5827.2010.00967.x; A. Bennett and V. Hayssen, "Measuring Cortisol in Hair and Saliva from Dogs: Coat Color and Pigment Differences," *Domestic Animal Endocrinology* 39 (2010): 171–80, doi:10.1016/j.domaniend.2010.04.003; T. Koyama, Y. Omata, and A. Saito, "Changes in Salivary Cortisol Concentrations during a 24-Hour Period in Dogs," *Hormone and Metabolic Research* 35 (2003): 355–57.
62. Dunbar, "Olfactory Preferences in Dogs."

CHAPTER FIVE

1. L. D. Mech, *The Wolf: The Ecology and Behavior of an Endangered Species* (Natural History Press, 1970), 389.
2. P. T. Iversen, conversation with author, February 27, 2013.
3. V. Fenton, "The Use of Dogs in Search, Rescue and Recovery," *Journal of Wilderness Medicine* 3 (1992): 292–300, doi:10.1580/0953-9859-3.3.292; S. Bulanda, *Ready! The Training of the Search and Rescue Dog* (Doral Publishing, 1994).
4. S. Charleson, *Scent of the Missing: Love and Partnership with a Search-and-Rescue Dog* (Mariner Books, Houghton Mifflin Harcourt, 2010); K. E. Jones et al., "Search-and-Rescue Dogs: An Overview for Veterinarians," *Journal of*

the American Veterinary Medical Association 225 (2004): 854–60, doi:10.2460/javma.2004.225.854.

5. S. Bulanda with L. Bulanda, *Ready! Training the Search and Rescue Dog*, 2nd ed. (Kennel Club Books, 2010). How to train your dog to track people or animals is more fully described in many other books. M. Hagstrom, *Nosarbete. En bok om spår & uppletande* (Skogsborgs Gård, 2007); I. Handegård, *Gå spor. Fra nybegynner til ekspert* (Pegasus forlag as, 2013); L. Fält et al., *Spårhunden och lukterna* (SWDI forlag, 2011); in English: *Tracking Dogs: Scents and Skills*, https://webshop.spesialsok.no/home/10-tracking-dogs-scent-and-skills.html; T. M. Hansen, *På sporet: en innføring i hvordan en kan lære hunden å gå spor. En enkel beskrivelse for folk flest* (Hansen, T. M., 2010); A. L. Mellin, *Sök & räddningshunden* (Mellin, A. L., 2012); S. Nordin, *Spårhundsboken* (Naturia Forlag AB, 1995); I. Sjösten and R. Sjösten, *Nya spårhunden* (SRI Publication AB, 2002); S. Järverud, *Din hund söker* (Brukshundservice Sverige AB, 2002).
6. Jones et al., "Search-and-Rescue Dogs."
7. J. Schettler, *K-9 Trailing: The Straightest Path* (Alpine Publications, 2011).
8. Ibid.
9. J. Schettler, *Red Dog Rising* (Alpine Publications, 2009); Schettler, *K-9 Trailing*.
10. M. T. Miller, *Search and Rescue Dogs: Dog Tales—True Stories about Amazing Dogs* (New York, 2007).
11. A. Bye, *Den store hundeboken* (Børrehaug & Co. Kunstforlag, 1963).
12. L. M. Harvey and J. W. Harvey, "Reliability of Bloodhounds in Criminal Investigations," *Journal of Forensic Sciences* 48 (2003): 811–16.
13. J. J. Ensminger, *Police and Military Dogs* (CRC Press, 2012).
14. Schettler, *Red Dog Rising*; M. Weisbord and K. Kachanoff, *Dogs with Jobs* (Pocket Books, 2000).
15. L. Rogak, *The Dogs of War: The Courage, Love, and Loyalty of Military Working Dogs* (Thomas Dunne Books, 2011).
16. Harvey and Harvey, "Reliability of Bloodhounds in Criminal Investigations."
17. A. Goldblatt, I. Gazit, and J. Terkel, "Olfaction and Explosives Detector Dogs," in *Canine Ergonomics: The Science of Working Dogs*, ed. W. S. Helton (CRC Press, 2009), 135–74.
18. M. McCarthy and P. Ahern, *In Search of the Missing: Working with Search and Rescue Dogs* (Mercier Press, 2011).
19. Schettler, *Red Dog Rising*; Schettler, *K-9 Trailing*.
20. A. Ferworn, *Canine Augmentation Technology for Urban Search and Rescue* (Taylor and Francis, 2009).
21. W. T. Chiu et al., "A Survey of International Urban Search-and-Rescue Teams Following the Ji Ji Earthquake," *Disasters* 26 (2002): 85–94, doi:10.1111/1467-7717.00193; Bulanda with Bulanda, *Ready!*, 2nd ed.; S. M. Hammond, *Training the Disaster Search Dog* (Dogwise, 2006).
22. American Rescue Dog Association, *Search and Rescue Dogs: Training the K-9 Hero*, 2nd ed. (Howell Book House, Wiley Publishing, 2002); C. M. Otto et al., "Medical

and Behavioral Surveillance of Dogs Deployed to the World Trade Center and the Pentagon from October 2001 to June 2002," *Journal of the American Veterinary Medical Association* 225 (2004): 861–67; N. K. Bauer, *Dog Heroes of September 11th: A Tribute to America's Search and Rescue Dogs* (Kennel Club Books, 2011).

23. Otto et al., "Medical and Behavioral Surveillance of Dogs Deployed to the World Trade Center and the Pentagon."
24. American Rescue Dog Association, *Search and Rescue Dogs*.
25. McCarthy and Ahern, *In Search of the Missing*.
26. D. Grandjean and F. Haymann, *Hunde-encyklopedi* (RK Grafisk, 2010).
27. L. Lit and C. A. Crawford, "Effects of Training Paradigms on Search Dog Performance," *Applied Animal Behaviour Science* 98 (2006): 277–92, doi:10.1016/j.applanim.2005.08.022.
28. Bulanda with Bulanda, *Ready!*, 2nd ed.
29. R. Huo et al., "The Trapped Human Experiment," *Journal of Breath Research* 5 (2011), doi:10.1088/1752-7155/5/4/046006.
30. Ibid.
31. Bulanda, *Ready!*; Ferworn, *Canine Augmentation Technology for Urban Search and Rescue*.
32. Network-Centric Applied Research Team, http://ncart.scs.ryerson.ca/research/cat/.
33. J. Tran et al., "Enhancing Canine Search," IEEE Systems of Systems Engineering Conference (SoSE '08) (Monterey, CA, June 2–5, 2008); J. Tran, M. Gerdzhev, and A. Ferworn, "Continuing Progress in Augmenting Urban Search and Rescue Dogs," Sixth International Wireless Communications and Mobile Computing Conference (IWCMC 2010) (June 28–July 2, 2010, Caen, France).
34. D. J. Wrigglesworth et al., "Accuracy of the Use of Triaxial Accelerometry for Measuring Daily Activity as a Predictor of Daily Maintenance Energy Requirement in Healthy Adult Labrador Retrievers," *American Journal of Veterinary Research* 72 (2011): 1151–55; C. Ribeiro et al., "Canine Pose Estimation: A Computing for Public Safety Solution," *2009 Canadian Conference on Computer and Robot Vision* (2009): 37–44; C. Ribeiro et al., "Wireless Estimation of Canine Pose for Search and Rescue," IEEE Systems of Systems Engineering conference (SoSE '08) (Monterey, CA, June 2–5, 2008); Ferworn, *Canine Augmentation Technology for Urban Search and Rescue*; Ribeiro et al., "Canine Pose Estimation."
35. M. Boecker et al., "Prefrontal Brain Activation during Stop-Signal Response Inhibition: An Event-Related Functional Near-Infrared Spectroscopy Study," *Behavioural Brain Research* 176 (2007): 259–66, doi:10.1016/j.bbr.2006.10.009; W. S. Helton et al., "The Abbreviated Vigilance Task and Cerebral Hemodynamics," *Journal of Clinical and Experimental Neuropsychology* 29 (2007): 545–52, doi:10.1080/13803390600814757; M. J. Herrmann et al., "Brain Activation for Alertness Measured with Functional Near Infrared Spectroscopy

(fNIRS)," *Psychophysiology* 45 (2008): 480–86, doi:10.1111/j.1469-8986.2007.00633.x.

36. Ferworn, *Canine Augmentation Technology for Urban Search and Rescue.*
37. K. E. Michel and D. C. Brown, "Determination and Application of Cut Points for Accelerometer-Based Activity Counts of Activities with Differing Intensity in Pet Dogs," *American Journal of Veterinary Research* 72 (2011): 866–70.
38. J. Tran and A. Ferworn, "Bark Indication Detection and Release Algorithm for the Automatic Delivery of Packages by Dogs," Sixth International Wireless Communications and Mobile Computing Conference (IWCMC 2010) (June 28–July 2, 2010, Caen, France); J. Tran et al., "Canine-Assisted Robot Deployment for Urban Search and Rescue," Safety Security and Rescue Robotics (SSRR), 2010 IEEE International Workshop, July 26–30, 2010; M. Gerdzhev et al., "DEX: A Design for Canine-Delivered Marsupial Robot," Safety Security and Rescue Robotics (SSRR), 2010 IEEE International Workshop, July 26–30, 2010.
39. B. G. Deshpande, "Earthquakes, Animals and Man, Chapter III: Animal Response to Earthquakes," *Proceedings of the National Academies of Science: India* B52, no. 5 (1986): 585–618; H. Tributsch, *When the Snakes Awake: Animals and Earthquake Prediction* (MIT Press, 1988).
40. Deshpande, "Earthquakes, Animals and Man."
41. J. Blumberg, "A Brief History of the St. Bernard Rescue Dog," *Smithsonian.com* (January 1, 2008).
42. K. Askildt, *Søk. 1956–2006, Norske Redningshunder gjennom femti år* (Thorsrud AS, 2006).
43. Blumberg, "A Brief History of the St. Bernard Rescue Dog"; P. Burnett, *Avalanche! Hasty Search: The Care and Training of Avalanche Search and Rescue Dogs* (Doral Publishing, 2003).
44. Burnett, *Avalanche! Hasty Search.*
45. Bye, *Den store hundeboken.*
46. W. G. Syrotuck, *Scent and the Scenting Dog* (Arner Publishing, 2000).
47. Burnett, *Avalanche! Hasty Search.*
48. M. Falk, H. Brugger, and L. Adlerkastner, "Avalanche Survival Chances," *Nature* 368 (1994): 21, doi:10.1038/368021a0.
49. Askildt, *Søk. 1956–2006, Norske Redningshunder gjennom femti år.*
50. P. O. Torkildsen, "Savnet og ettersøkt- en studie om savnede personer på land i Norge og de søk som blir iverksatt for å finne dem," PHS-skriftserie, Politiskolen, Oslo 2 (2009).
51. Miller, *Search and Rescue Dogs.*
52. Askildt, *Søk. 1956–2006, Norske Redningshunder gjennom femti år.*
53. Burnett, *Avalanche! Hasty Search.*
54. Ibid.
55. Askildt, *Søk. 1956–2006, Norske Redningshunder gjennom femti år.*

CHAPTER SIX

1. H. Parker, e-mail message to author, January 8, 2014.
2. V. Ruusila and M. Pesonen, "Interspecific Cooperation in Human (*Homo sapiens*) Hunting: The Benefits of a Barking Dog (*Canis familiaris*), *Annales Zoologici Fennici* 41 (2004): 545–49.
3. J. M. Koster, "Hunting with Dogs in Nicaragua: An Optimal Foraging Approach," *Current Anthropology* 49 (2008): 935–44, doi:10.1086/592021.
4. D. Grandjean and F. Haymann, *Hunde-encyklopedi* (RK Grafisk, 2010).
5. N. Lescureux and J. D. C. Linnell, "Warring Brothers: The Complex Interactions between Wolves (*Canis lupus*) and Dogs (*Canis familiaris*) in a Conservation Context," *Biological Conservation* 171 (2014): 232–45, doi:http://dx.doi.org/10.1016/j.biocon.2014.01.032.
6. H. J. Huson, *Genetic Aspects of Performance in Working Dogs* (CAB International, 2012).
7. J. A. Ewald et al., "Fox-Hunting in England and Wales: Its Contribution to the Management of Woodland and Other Habitats," *Biodiversity and Conservation* 15 (2006): 4309–34, doi:10.1007/s10531-005-3739-z.
8. "Foxhunting: Will the Ban on Hunting Affect Fox Numbers?" *The Fox Website*, http://www.thefoxwebsite.org/foxhunting/huntban.
9. S. Christoffersson, *Jakt i Norden* (Cappelen Damm AS, 2008).
10. K. V. Pedersen and J. Unsgård, *Elghunder og elgjakt* (Tun Forlag, 2010); R. Andersen, A. Mysterud, and E. Lund, *Rådyret- det lille storviltet* (Naturforlaget, 2004); K. V. Pedersen, *Rådyrjakt med hund* (Tun Forlag, 2007); E. L. Meisingset, *Hjort og hjortejakt i Norge* (Naturforlaget, 2003); E. L. Meisingset, *Alt om hjort biologi, jakt, forvaltning* (Tun Forlag, 2008); S. Lier-Hansen, *Villrein & villreinjakt* (Landbruksforlaget, 1994).
11. Ruusila and Pesonen, "Interspecific Cooperation in Human (*Homo sapiens*) Hunting."
12. C. Godwin et al., "Contribution of Dogs to White-Tailed Deer Hunting Success," *Journal of Wildlife Management* 77 (2013): 290–96, doi:10.1002/jwmg.474.
13. H. C. Pedersen and D. H. Karlsen, *Alt om rypa, biologi, jakt, forvaltning* (Tun Forlag, 2007); S. S. Wangen and R. Sagen, *Rypejakt og jakten på den perfekte fuglehund* (Kom Forlag, 2006); J. Barikmo and H. C Pedersen,. *Harer og harejakt* (Gyldendal Fakta, 1997); J. C. Frøstrup, *Hare og harejakt* (Teknologisk forlag, 1996); Christoffersson, *Jakt i Norden.*; K. V. Pedersen, *Småvilt- jakt of fangst* (Tun Forlag AS, 2008).
14. Pedersen, *Småvilt- jakt of fangst.*
15. Christoffersson, *Jakt i Norden.*
16. F. Rosell and K. V. Pedersen, *Bever* (Landbruksforlaget, 1999).
17. M. Gackis and Z. Berzina, "Beaver Hunting Success Using Whole Family Hunt-Out Method," Sixth International Beaver Symposium (Ivanić Grad, Croatia,

September 17–20, 2012); M. Gackis, e-mail and conversations in Croatia during September 2012.

18. K. J. Gutzwiller, "Minimizing Dog-Induced Biases in Game Bird Research," *Wildlife Society Bulletin* 18 (1990): 351–56; Campagna et al., "Potential Semiochemical Molecules from Birds: A Practical and Comprehensive Compilation of the Last 20 Years Studies," *Chemical Senses* 37 (2012): 3–25, doi:10.1093/chemse/bjr067.
19. D. Müller-Schwarze, *Chemical Ecology of Vertebrates* (Cambridge University Press, 2006); M. R. Conover, *Predator-Prey Dynamics: The Role of Olfaction* (CRC Press, 2007); Campagna et al., "Potential Semiochemical Molecules from Birds."
20. Conover, *Predator-Prey Dynamics.*
21. W. G. Syrotuck, *Scent and the Scenting Dog* (Arner Publishing, 2000).
22. S. Bulanda, *Scenting in the Wind: Scentwork for Hunting Dogs* (Doral Publishing, 2002).
23. Conover, *Predator-Prey Dynamics.*
24. G. W. Gabrielsen, A. S. Blix, and H. Ursin, "Orienting and Freezing Responses in Incubating Ptarmigan Hens," *Physiology & Behavior* 34 (1985): 925–34, doi:10.1016/0031-9384(85)90015-0; J. B. Steen, G. W. Gabrielsen, and J. W. Kanwisher, "Physiological Aspects of Freezing Behavior in Willow Ptarmigan Hens," *Acta Physiologica Scandinavica* 134 (1988): 299–304, doi:10.1111/j.1748-1716.1988.tb08493.x.
25. Conover, *Predator-Prey Dynamics.*
26. Syrotuck, *Scent and the Scenting Dog*; Conover, *Predator-Prey Dynamics.*
27. Müller-Schwarze, *Chemical Ecology of Vertebrates.*
28. Andersen, Mysterud, and Lund, *Rådyret- det lille storviltet.*
29. Müller-Schwarze, *Chemical Ecology of Vertebrates*; Campagna et al., "Potential Semiochemical Molecules from Birds."
30. R. S. Cook et al., "Mortality of Young White-Tailed Deer Fawns in South Texas," *Journal of Wildlife Management* 35 (1971): 47–56, doi:10.2307/3799870.
31. Müller-Schwarze, *Chemical Ecology of Vertebrates*; Conover, *Predator-Prey Dynamics.*
32. D. A. S. Woollett, A. Hurt, and N. L. Richards, "The Current and Future Roles of Free-Ranging Detection Dogs in Conservation Efforts," in *Free-Ranging Dogs and Wildlife Conservation*, ed. M. E. Gompper (Oxford University Press, 2014), 239–64.
33. F. Rosell et al., "Predator-Naive Brown Trout (*Salmo trutta*) Show Antipredator Behaviours to Scent from an Introduced Piscivorous Mammalian Predator Fed Conspecifics," *Ethology* 119 (2013): 303–8, doi:10.1111/eth.12065.
34. Syrotuck, *Scent and the Scenting Dog*; Conover, *Predator-Prey Dynamics.*
35. J. A. Shivik, "Odor-Adsorptive Clothing, Environmental Factors, and Search-Dog Ability," *Wildlife Society Bulletin* 30 (2002): 721–27.
36. Syrotuck, *Scent and the Scenting Dog*; Conover, *Predator-Prey Dynamics.*
37. S. K. Wasser et al., "Scat Detection Dogs in Wildlife Research and Management: Application to Grizzly and Black Bears in the Yellowhead Ecosystem, Alberta,

Canada," *Canadian Journal of Zoology/Revue canadienne de zoologie* 82 (2004): 475–92, doi:10.1139/z04-020.

38. V. Geist, "Adaptive Behavioral Strategies," in *North American Elk: Ecology and Management*, ed. D. E. Toweill and J. W. Thomas (Smithsonian Institute Press, 2002), 389–433.
39. Conover, *Predator-Prey Dynamics*.
40. Syrotuck, *Scent and the Scenting Dog*.
41. O. Reitan, e-mail message to author, February 16, 2014.
42. Bulanda, *Scenting in the Wind*.
43. O. Reitan, e-mail message to author, February 16, 2014.
44. Conover, *Predator-Prey Dynamics*.
45. Ibid.
46. Gutzwiller, "Minimizing Dog-Induced Biases in Game Bird Research."
47. Bulanda, *Scenting in the Wind*.
48. S. M. Stejskal, *Death, Decomposition, and Detector Dogs* (CRC Press, 2013).
49. Bulanda, *Scenting in the Wind*.
50. Conover, *Predator-Prey Dynamics*.
51. I. Hanssen-Bauer, e-mail message to author, March 30, 2014.
52. Conover, *Predator-Prey Dynamics*.
53. Ibid.
54. Ibid.
55. Ibid.
56. Syrotuck, *Scent and the Scenting Dog*.
57. Ibid.
58. Müller-Schwarze, *Chemical Ecology of Vertebrates*.
59. Conover, *Predator-Prey Dynamics*.
60. Müller-Schwarze, *Chemical Ecology of Vertebrates*.
61. Conover, *Predator-Prey Dynamics*.
62. Ibid.
63. I. Hanssen-Bauer, e-mail message to author, March 30, 2014.
64. Conover, *Predator-Prey Dynamics*.
65. H. M. Budgett, *Hunting by Scent* (Charles Scribner's Sons, 1933).
66. Syrotuck, *Scent and the Scenting Dog*.
67. J. B. Steen and E. Wilsson, *Fuglehundens ABC & D* (Hundskolan i Sollefteå AB, 1993).
68. Syrotuck, *Scent and the Scenting Dog*.
69. I. Hanssen-Bauer, e-mail message to author, March 30, 2014.
70. J. Jeanneney, *Tracking Dogs for Finding Wounded Deer* (Teckel Time Inc., 2006); A. Buvik, *Ettersøkshunden. Trening og godkjenningsprøver* (Landbruksforlaget, 1993).
71. S. Stokke et al., *Kan vi stole på våre ettersøkshunder? En evaluering av godkjente ettersøkshunders sporingsevne på hjortevilt* (Norsk institutt for naturforskning, Trondheim, 2011).
72. Ibid.

73. G. K. Moen et al., "Behaviour of Solitary Adult Scandinavian Brown Bears (*Ursus arctos*) When Approached by Humans on Foot," *PLOS ONE* 7 (2012), doi:10.1371/journal.pone.0031699.
74. J. E. Swenson et al., "Interactions between Brown Bear and Humans in Scandinavia," *Biosphere Conservation* 2 (1999): 1–9.
75. S. Stokke et al., "Märkbart många björnar skadeskjuts," *Våre Rovdyr* 2 (2008): 18–20.
76. S. Vang et al., *Sporing av bjørn. En empirisk studie av ettersøksekvipasjer på bjørn sommeren 2007 og 2008* (Universitet for miljø- og biovitenskap Ås, 2009); S. Vang et al., "Ettersøksekvipasjene ikke gode nok," *Jakt og Fiske* 1–2 (2011): 64–68.
77. S. Vang et al., *Sporing av bjørn*; Vang et al., "Ettersøksekvipasjene ikke gode nok."
78. S. Vang, e-mail message to author, June 18, 2014.
79. B. Forkman et al., "Dogs Hunting Bears—What Do They Really Hunt?" Third Canine Science Forum (Barcelona, July 25–27, 2012).
80. Stokke et al., *Kan vi stole på våre ettersøkshunder?*; Vang et al., *Sporing av bjørn*; S. Vang and G. K. Moen, *Kan ettersøksekvipasjer spore bjørn: en empirisk studie av norsk og svensk ettersøksekvipasjer på bjørn* (UMB, ÅS, Semesteroppgave. Det skandinaviske bjørneprosjektet, 2007).

CHAPTER SEVEN

1. A. Bye, *Den store hundeboken* (Børrehaug & Co. Kunstforlag, 1963).
2. G. A. A. Schoon and R. Haak, *K9 Suspect Discrimination* (Detseling Enterprises Ltd, 2002).
3. S. S. Ozcan et al., "Utilization of Police Dogs: A Turkish Perspective," *Policing: An International Journal of Police Strategies & Management* 32 (2009): 226–37, doi:10.1108/13639510910958154.
4. N. Pemberton, "'Bloodhounds as Detectives': Dogs, Slum Stench and Late-Victorian Murder Investigation," *Cultural & Social History* 10 (2013): 69–91, doi:10.2752/147800413X13515292098197; N. Pemberton, "Hounding Holmes: Arthur Conan Doyle, Bloodhounds and Sleuthing in the Late-Victorian Imagination," *Journal of Victorian Culture* 17 (2012): 454–67, doi:10.1080/13555502.2012.737099.
5. N. Allsopp, *K9 Cops: Police Dogs of the World* (Big Sky Publishing, 2012); B. Barbour, *Review of the Police Powers (Drug Detection Dogs) Act 2001* (NSW Ombudsman, Sydney, 2006).
6. Allsopp, *K9 Cops*.
7. R. A. Stochkham, D. L. Slavin, and W. Kift, "Survivability of Human Scent," *Forensic Science Communications* 6 (2004): 1–10.
8. N. Lorenzo et al., "Laboratory and Field Experiments Used to Identify *Canis lupus* var. *familiaris* Active Odor Signature Chemicals from Drugs, Explosives, and Humans," *Analytical and Bioanalytical Chemistry* 376 (2003): 1212–24, doi:10.1007/s00216-003-2018-7.
9. D. Grandjean and F. Haymann, *Hunde-encyklopedi* (RK Grafisk, 2010).

10. J. M. Slabbert and J. S. J. Odendaal, "Early Prediction of Adult Police Dog Efficiency: A Longitudinal Study." *Applied Animal Behaviour Science* 64 (1999): 269–88, doi:10.1016/s0168-1591(99)00038-6.
11. P. Jensen, *Hundens språk och tankar* (Natur & Kultur, 2012); C. Diederich and J. M. Giffroy, "Behavioural Testing in Dogs: A Review of Methodology in Search for Standardisation," *Applied Animal Behaviour Science* 97 (2006): 51–72, doi:10.1016/j.applanim.2005.11.018; L. Asher et al., "A Standardized Behavior Test for Potential Guide Dog Puppies: Methods and Association with Subsequent Success in Guide Dog Training," *Journal of Veterinary Behavior: Clinical Applications and Research* 8 (2013): 431–38, doi:10.1016/j.jveb.2013.08.004; K. Tiira and H. Lohi, "Reliability and Validity of a Questionnaire Survey in Canine Anxiety Research," *Applied Animal Behaviour Science* 155 (2014): 82–92, doi:http://dx.doi.org/10.1016/j.applanim.2014.03.007; P. Foyer et al., "Behaviour and Experiences of Dogs during the First Year of Life Predict the Outcome in a Later Temperament Test," *Applied Animal Behaviour Science* 155 (2014): 93–100.
12. Slabbert and Odendaal, "Early Prediction of Adult Police Dog Efficiency."
13. www.volhard.com.
14. G. A. A. Schoon, A. M. Curran, and K. G. Furton, *Odor Biometrics* (Springer Science + Business Media, 2009); T. Jezierski, M. Walczak, and A. Gorecka, "Information-Seeking Behaviour of Sniffer Dogs during Match-to-Sample Training in the Scent Lineup," *Polish Psychological Bulletin* 39 (2008): 71–80; T. Jezierski et al., "Operant Conditioning of Dogs (*Canis familiaris*) for Identification of Humans Using Scent Lineup," *Animal Science Papers and Reports* 28 (2010): 81–93.
15. J. Kaldenbach, *K9 Scent Detection: My Favorite Judge Lives in a Kennel* (Brush Education, 1998).
16. S. Bulanda with L. Bulanda, *Ready! Training the Search and Rescue Dog*, 2nd ed. (Kennel Club Books, 2010).
17. J. J. Ensminger, *Police and Military Dogs* (CRC Press, 2012).
18. Schoon and Haak, *K9 Suspect Discrimination*.
19. Ibid.; G. A. A. Schoon, "Scent Identification Line-Ups Using Trained Dogs in the Netherlands," *Problems of Forensic Sciences* 47 (2001): 175–83.
20. T. Tomaszewski and P. Girdwoyn, "Scent Identification Evidence in Jurisdiction (Drawing on the Example of Judicial Practice in Poland)," *Forensic Science International* 162 (2006): 191–95, doi:10.1016/j.forsciint.2006.06.017; R. H. Settle et al., "Human Scent Matching Using Specially Trained Dogs," *Animal Behaviour* 48 (1994): 1443–48, doi:10.1006/anbe.1994.1380; Stochkham, Slavin, and Kift, "Survivability of Human Scent"; Kaldenbach, *K9 Scent Detection*; Schoon, "Scent Identification Line-Ups Using Trained Dogs in the Netherlands."
21. G. A. A. Schoon, "The Effect of the Ageing of Crime Scene Objects on the Results of Scent Identification Line-Ups Using Trained Dogs," *Forensic Science International* 147 (2005): 43–47, doi:10.1016/j.forsciint.2004.04.080.

22. Ibid.
23. S. H. Jenkins, "Can Police Dogs Identify Criminal Suspects by Smell? Using Experiments to Test Hypotheses about Animal Behavior," in *How Science Works: Evaluating Evidence in Biology and Medicine* (Oxford University Press, 2004), 36–52.
24. Kaldenbach, *K9 Scent Detection*; R. H. Settle et al., "Human Scent Matching Using Specially Trained Dogs"; Schoon, "The Effect of the Ageing of Crime Scene Objects"; G. A. A. Schoon and J. C. Debruin, "The Ability of Dogs to Recognize and Cross-Match Human Odors," *Forensic Science International* 69 (1994): 111–18, doi:10.1016/0379-0738(94)90247-x; G. A. A. Schoon, "Scent Identification Line-ups by Dogs (*Canis familiaris*): Experimental Design and Forensic Application," *Applied Animal Behaviour Science* 49 (1996): 257–67, doi:10.1016/0168-1591(95)00656-7.
25. A. Rebmann, E. David, and M. Sorg, *Cadaver Dog Handbook: Forensic Training and Tactics for the Recovery of Human Remains* (CRC Press, 2000).
26. W. C. Rodriguez and W. M. Bass, "Decomposition of Buried Bodies and Methods That May Aid in Their Location," *Journal of Forensic Sciences* 30 (1985): 836–52.
27. P. S. Martin et al., "Dog versus Machine: Exploring the Utility of Cadaver Dogs and Ground-Penetrating Radar in Locating Human Burials at Historical Archeological Sites," Geological Society of America, Southeastern Section, 61st Annual Meeting April 1–2, 2012.
28. A. F. Migala and S. E. Brown, "Use of Human Remains Detection Dogs for Wide Area Search after Wildfire: A New Experience for Texas Task Force 1 Search and Rescue Resources," *Wilderness & Environmental Medicine* 23 (2012): 337–42.
29. Rebmann, David, and Sorg, *Cadaver Dog Handbook*; A. E. Lasseter et al., "Cadaver Dog and Handler Team Capabilities in the Recovery of Buried Human Remains in the Southeastern United States," *Journal of Forensic Sciences* 48 (2003): 617–21.
30. Rebmann, David, and Sorg, *Cadaver Dog Handbook*.
31. M. Korneliussen, "Likhunders markeringer ved molekylære søk som bevismiddel i straffesaker" (master's thesis, University of Tromsø, Tromsø, Norway, 2007).
32. P. A. Sødal, e-mail message to author, June 17, 2014; P. A. Prada, A. M. Curran, and K. G. Furton, "Characteristic Human Scent Compounds Trapped on Natural and Synthetic Fabrics as Analyzed by SPME-GC/MS," *Journal of Forensic Science & Criminology* 1 (2014).
33. P. A. Prada, A. M. Curran, and K. G. Furton, "Characteristic Human Scent Compounds Trapped on Natural and Synthetic Fabrics as Analyzed by SPME-GC/MS," *Journal of Forensic Science & Criminology* 1 (2014).
34. T. Jezierski, e-mail message to author, June 16, 2014.
35. S. M. Stejskal, *Death, Decomposition, and Detector Dogs* (CRC Press, 2013).
36. D. L. France et al., "A Multidisciplinary Approach to the Detection of Clandestine Graves," *Journal of Forensic Sciences* 37 (1992): 1445–58.

37. Stejskal, *Death, Decomposition, and Detector Dogs*.
38. M. Statheropoulos, C. Spiliopouiou, and A. Agapiou, "A Study of Volatile Organic Compounds Evolved from the Decaying Human Body," *Forensic Science International* 153 (2005): 147–55, doi:10.1016/j.forsciint.2004.08.015.
39. A. A. Vass et al., "Odor Analysis of Decomposing Buried Human Remains," *Journal of Forensic Sciences* 53 (2008): 384–91, doi:10.1111/j.1556-4029.2008.00680.x; D. Page, "Labrador: New Alpha Dog in Human Remains Detection?" *Forensic Magazine* 5 (2010): 33–40.
40. Korneliussen, "Likhunders markeringer ved molekylære søk som bevismiddel i straffesaker."
41. Lorenzo et al., "Laboratory and Field Experiments Used to Identify *Canis lupus* var. *familiaris* Active Odor Signature Chemicals from Drugs, Explosives, and Humans"; Vass et al., "Odor Analysis of Decomposing Buried Human Remains."
42. Page, "Labrador: New Alpha Dog in Human Remains Detection?"
43. M. E. Cablk, E. E. Szelagowski, and J. C. Sagebiel, "Characterization of the Volatile Organic Compounds Present in the Headspace of Decomposing Animal Remains, and Compared With Human Remains," *Forensic Science International* 220 (2012): 118–25, doi:10.1016/j.forsciint.2012.02.007.
44. A. Grande, e-mail message to author, October 2013.
45. M. Koenig, e-mail message to author, October 8, 2013.
46. B. Jones, e-mail message to author, October 8, 2013.
47. N. Lorenzo et al., "Laboratory and Field Experiments Used to Identify *Canis lupus* var. *familiaris* Active Odor Signature Chemicals from Drugs, Explosives, and Humans"; U. Ellervik, *Ond kemi. Berättelser om människor, mord och molekyler* (Fri tanke forlag, 2011).
48. L. Oesterhelweg et al., "Cadaver Dogs: A Study on Detection of Contaminated Carpet Squares," *Forensic Science International* 174 (2008): 35–39, doi:10.1016/j.forsciint.2007.02.031; M. Cooke, N. Leeves, and C. White, "Time Profile of Putrescine, Cadaverine, Indole and Skatole in Human Saliva," *Archives of Oral Biology* 48 (2003): 323–27, doi:10.1016/s0003-9969(03)00015-3.
49. S. Stadler et al., "Analysis of Synthetic Canine Training Aids by Comprehensive Two-Dimensional Gas Chromatography-Time of Flight Mass Spectrometry," *Journal of Chromatography A* 1255 (2012): 202–6, doi:10.1016/j.chroma.2012.04.001.
50. L. E. DeGreeff, B. Weakley-Jones, and K. G. Furton, "Creation of Training Aids for Human Remains Detection Canines Utilizing a Non-Contact, Dynamic Airflow Volatile Concentration Technique," *Forensic Science International* 217 (2012): 32–38, doi:10.1016/j.forsciint.2011.09.023.
51. D. T. Hudson-Holness and K. G. Furton, "Comparison between Human Scent Compounds Collected on Cotton and Cotton Blend Materials for SPME-GC/MS Analysis," *Journal of Forensic Research* 1 (2010), doi:10.4172/2157-7145.1000101.
52. Oesterhelweg et al., "Cadaver Dogs."
53. S. Bulanda, e-mail message to author, March 19, 2013.

54. C. Judah, *Search and Rescue Dogs: Water Search: Finding Drowned Persons* (Coastal Books, 2011).
55. T. Osterkamp, "K9 Water Searches: Scent and Scent Transport Considerations," *Journal of Forensic Sciences* 56 (2011): 907–12, doi:10.1111/j.1556-4029.2011.01773.x.
56. K. I. Aasheim, "Hund som arbeidsredskap i brannvesenet," *Brannmannen* 61 (2006): 42–43.
57. K. C. Catania, J. F. Hare, and K. L. Campbell, "Water Shrews Detect Movement, Shape, and Smell to Find Prey Underwater," *Proceedings of the National Academy of Sciences of the United States of America* 105 (2008): 571–76, doi:10.1073/pnas.0709534104.
58. Bulanda with Bulanda, *Ready!*, 2nd ed.
59. D. Müller-Schwarze, *Chemical Ecology of Vertebrates* (Cambridge University Press, 2006).
60. Bulanda with Bulanda, *Ready!*, 2nd ed.
61. Røde Kors Førstehjelp, http://www.rodekorsforstehjelp.no/default.aspx?articleid=115&pageId=59&news=1.
62. J. C. Judah, *Buzzards and Butterflies: Human Remains Detection Dogs* (Coastal Books, 2008).
63. Bulanda with Bulanda, *Ready!*, 2nd ed.
64. H. Graham and J. Graham, "Taking Back from the River," *Response* (September/October 1987): 34–37.
65. M. T. Miller, *Search and Rescue Dogs: Dog Tales—True Stories about Amazing Dogs* (New York, 2007).
66. P. A. Sødal, e-mail message to author, October 2013.
67. J. S. Brown et al., "Applicability of Emanating Volatile Organic Compounds from Various Forensic Specimens for Individual Differentiation," *Forensic Science International* 226 (2013): 173–82, doi:10.1016/j.forsciint.2013.01.008.
68. S. Newbery and J. P. Cassella, "A Study of the Use of Cadaver Dogs for Blood Scent Detection in Criminal Investigations," Abstract, FIRN Midlands Regional Student Forensic Science Conference (2008).
69. M. M. Bouzga and A. G. Skalleberg, "Hundeprosjektet," *Bevis* 3 (2013): 23–25; A. G. Skalleberg, "Hundeprosjektet," *Bevis* 1 (2012): 8–10; A. G. Skalleberg, "Hundeprosjektet," *Bevis* 2 (2013): 14–16.
70. M. E. Cablk and J. C. Sagebiel, "Field Capability of Dogs to Locate Individual Human Teeth," *Journal of Forensic Sciences* 56 (2011): 1018–24, doi:10.1111/j.1556-4029.2011.01785.x.
71. Allsopp, *K9 Cops*.
72. USDA, "K9 Friendly Fire Prevention," http://www.na.fs.fed.us/ss/12/K9FriendlyFirePreventionSS_121107.pdf.
73. R. A. Strobel and R. Noll, "Pilot Project Arson Accelerant Detector Dog Program: The Forensic Science Laboratory's Role," *Proceedings of the Canine Accelerant Detection Training Seminar* (Connecticut, USA, August 1988); W. Smith et al.,

"Pilot Project Arson Accelerant Detector Dog Program," *Proceedings of the Canine Accelerant Detection Training Seminar* (Connecticut, USA, August 1988).

74. D. M. Gialamas, "Enhancement of Fire Scene Investigations Using Accelerant Detection Canines," *Science & Justice* 36 (1996): 51–54, doi:10.1016/s1355-0306(96)72555-8.
75. R. Tindall and K. Lothridge, "An Evaluation of 42 Accelerant Detection Canine Teams," *Journal of Forensic Sciences* 40 (1995): 561–64.
76. M. Weisbord and K. Kachanoff, *Dogs with Jobs* (Pocket Books, 2000).
77. Gialamas, "Enhancement of Fire Scene Investigations Using Accelerant Detection Canines."
78. S. R. Katz and C. R. Midkiff, "Unconfirmed Canine Accelerant Detection: A Reliability Issue in Court," *Journal of Forensic Sciences* 43 (1998): 329–33.
79. Weisbord and Kachanoff, *Dogs with Jobs*.
80. M. E. Kurz et al., "Evaluation of Canines for Accelerant Detection at Fire Scenes," *Journal of Forensic Sciences* 39 (1994): 1528–36; P. M. Sandercock, "Fire Investigation and Ignitable Liquid Residue Analysis: A Review, 2001–2007," *Forensic Science International* 176 (2008): 93–110, doi:10.1016/j.forsciint.2007.09.004.
81. State Farm, http://www.statefarm.com/aboutus/_pressreleases/2012/new-arson-dog-nc.asp.
82. O. M. Mortvedt, "For første gang i Norge- brannhund på åstedet," *Politiforum* 2 (2003): 8.
83. Gialamas, "Enhancement of Fire Scene Investigations Using Accelerant Detection Canines"; S. Jacobs, "K-9s Prove Their Worth," *Fire & Arson Investigator* 43 (1993): 50.
84. P. A. Sødal, e-mail message to author, October 2013.
85. Sandercock, "Fire Investigation and Ignitable Liquid Residue Analysis: A Review, 2001–2007."
86. M. Nowlan et al., "Use of a Solid Absorbent and an Accelerant Detection Canine for the Detection of Ignitable Liquids Burned in a Structure Fire," *Journal of Forensic Sciences* 52 (2007): 643–48, doi:10.1111/j.1556-4029.2007.00408.x.
87. D. J. Tranthim-Fryer and J. D. DeHaan, "Canine Accelerant Detectors and Problems with Carpet Pyrolysis Products," *Science & Justice* 37 (1997): 39–46, doi:10.1016/s1355-0306(97)72139-7.
88. P. A. Sødal, e-mail message to author, October 2013; The manual *United Kingdom Fire Investigation Dog and Handler Teams: Guide to the Best Practice* (Office of the Deputy Prime Minister, London, January 2004); Scientific Working Group on Dog and Orthogonal Detector Guidelines (SWGDOG) has published some guidelines for dog teams: see http://swgdog.fiu.edu/.
89. A. Hallgren, *Smoke Alarm Training for Your Dog* (Hallwig Publishing, 2002); translation of A. Hallgren, *Livräddaren på fyra ben* (Mälaröbörsen Forlag AB, 2002).
90. I. Jalakas, *Den nyttiga nosen* (Bilda Forlag, 2000).
91. Ibid.

92. Ellervik, *Ond kemi. Berättelser om människor, mord och molekyler*.
93. Aasheim, "Hund som arbeidsredskap i brannvesenet."

CHAPTER EIGHT

1. M. Weisbord and K. Kachanoff, *Dogs with Jobs* (Pocket Books, 2000).
2. J. Ashley, S. Billinge, and C. Hemmens, "Who Let the Dogs Out? Drug Dogs in Court," *Criminal Justice Studies* 20 (2007): 177–96.
3. R. T. Rolak, "Use of Police Canine Units in Narcotic Searches of Vehicles" (School of Police Staff and Command Skip Lawver- Eastern Michigan University, 2000), 1–28, http://citeseerx.ist.psu.edu/viewdoc/download?doi=10.1.1.537.1664&rep=rep1&type=pdf.
4. D. Grandjean and F. Haymann, *Hunde-encyklopedi* (RK Grafisk, 2010).
5. P. W. Zeno, "Going to the Dogs," *AAA Going Places Magazine* (May–June 1998): 20–22.
6. Grandjean and Haymann, *Hunde-encyklopedi*.
7. USDA, "USDA's Detector Dogs: Protecting American Agriculture," *Animal Plant Health Inspection Service, Miscellaneous Publication No. 1539* (1996): 1–13.
8. J. B. Stamper, *Eco Dogs* (Bearport Publishing, 2011); P. MacKay, "Conservation Dogs Work for Wildlife," *The Bark*, http://thebark.com/content/conservation-dogs-work-wildlife.
9. R. I. Orenstein, *Ivory, Horn and Blood: Behind the Elephant and Rhinoceros Poaching Crisis* (Firefly Books, 2013).
10. "Yemen Acts to Halt Rhino Horn Daggers; Scientific Tests Fail to Show Rhino Horn Effective as Medicine," *Environmentalist* 3 (1983): 153.
11. E. Hardcastle, "African Rhino Poaching Hits Record on Cancer Claim," *Reuters*, October 16, 2012.
12. R. Starkey, e-mail message to author, October 2013; T. Kermeliotis, "Poaching Stinks . . . and Now Dogs Are Sniffing It Out," CNN.com, September 7, 2013, http://edition.cnn.com/2013/09/25/world/africa/poaching-stinks-dogs-sniffing/index.html.
13. D. B. Chhabra, e-mail message from Manager of Communications, TRAFFIC India, April 4, 2013; TRAFFIC, The Wildlife Trade Monitoring Network, http://www.traffic.org/; WWF-India, http://www.wwfindia.org/.
14. C. Scheiner, "Time of the Es'scent: The Fourth Amendment, Canine Olfaction, and Vehicle Stops," *Florida Bar Journal* 76 (2002): 26–32.
15. T. English, *The Quiet Americans: A History of Military Working Dogs* (Office of History, 37th Training Wing, Lackland AFB, 2000).
16. Ibid.
17. J. J. Ensminger, *Police and Military Dogs* (CRC Press, 2012).
18. W. S. Helton, ed., *Canine Ergonomics: The Science of Working Dogs* (CRC Press, 2009).
19. R. Aaron and P. Lewis, "Cocaine Residues on Money," *Crime Laboratory Digest* 14 (1987): 18; K. G. Furton et al., "Identification of Odor Signature Chemicals in

Cocaine Using Solid-Phase Microextraction-Gas Chromatography and Detector-Dog Response to Isolated Compounds Spiked on US Paper Currency," *Journal of Chromatographic Science* 40 (2002): 147–55; K. G. Furton et al., "Odor Signature of Cocaine Analyzed by GC/MS and Threshold Levels of Detection for Drug Detection Canines," *Current Topics in Forensic Science: Proceedings of the 14th International Association of Forensic Sciences* 2 (1997): 329–32.

20. C. Remsberg, *Tactics for Criminal Patrol: Vehicle Stops, Drug Discovery and Officer Survival* (Calibre Press, 1995).
21. B. Handwerk, "'Detector Dogs' Sniff Out Smugglers for U.S. Customs," *National Geographic News*, July 12, 2002.
22. T. Knapperholen, e-mail message to author, November 1, 2013.
23. J. Bradshaw, *In Defence of Dogs* (Penguin, 2011).
24. T. Jezierski et al., "Efficacy of Drug Detection by Fully-Trained Police Dogs Varies by Breed, Training Level, Type of Drug and Search Environment," *Forensic Science International* 237 (2014): 112–18, doi:http://dx.doi.org/10.1016/j.forsciint.2014.01.013.
25. M. A. Spoto, "Cell Phone-Sniffing Dogs Used to Crack Down on N.J. Prisoners Orchestrating Crime from Behind Bars," NJ.com, December 12, 2011, http://www.nj.com/news/index.ssf/2011/12/cell_phone-sniffing_dogs_used.html.
26. Grandjean and Haymann, *Hunde-encyklopedi*.
27. S. J. Orr, "Flo and Lucky, MPAA's DVD Sniffing Dogs," NJ.com May 14, 2008, http://blog.nj.com/digitallife/2008/05/theyre_not_exactly_film_critic.html.

CHAPTER NINE

1. L. Rogak, *The Dogs of War: The Courage, Love, and Loyalty of Military Working Dogs* (Thomas Dunne Books, 2011).
2. A. Bye, *Den store hundeboken* (Børrehaug & Co. Kunstforlag, 1963).
3. S. Coren, *The Intelligence of Dogs* (Bantam Books, 1994).
4. C. Hebard, "Use of Search and Rescue Dogs," *Journal of the American Veterinary Medical Association* 203 (1993): 999–1001.
5. N. Allsopp, *Cry Havoc: The History of War Dogs* (New Holland Publishers, 2011).
6. Bye, *Den store hundeboken*.
7. V. Fenton, "The Use of Dogs in Search, Rescue and Recovery," *Journal of Wilderness Medicine* 3 (1992): 292–300, doi:10.1580/0953-9859-3.3.292.
8. W. G. Syrotuck, *Scent and the Scenting Dog* (Arner Publishing, 2000).
9. D. Grandjean and F. Haymann, *Hunde-encyklopedi* (RK Grafisk, 2010).
10. Allsopp, *Cry Havoc*.
11. Ibid.
12. T. English, *The Quiet Americans: A History of Military Working Dogs* (Office of History, 37th Training Wing, Lackland AFB, 2000).
13. Allsopp, *Cry Havoc*.
14. Bye, *Den store hundeboken*.

15. Rogak, *The Dogs of War*.
16. Allsopp, *Cry Havoc*.
17. A. Goldblatt, I. Gazit, and J. Terkel, "Olfaction and Explosives Detector Dogs," in *Canine Ergonomics: The Science of Working Dogs*, ed. W. S. Helton (CRC Press, 2009), 135–74.
18. K. G. Furton and L. J. Myers, "The Scientific Foundation and Efficacy of the Use of Canines as Chemical Detectors for Explosives," *Talanta* 54 (2001): 487–500, doi:10.1016/s0039-9140(00)00546-4.
19. I. Gazit and J. Terkel, "Domination of Olfaction over Vision in Explosives Detection by Dogs," *Applied Animal Behaviour Science* 82 (2003): 65–73, doi:10.1016/s0168-1591(03)00051-0.
20. Furton and Myers, "The Scientific Foundation and Efficacy of the Use of Canines as Chemical Detectors for Explosives"; Grandjean and Haymann, *Hundeencyklopedi*.
21. Y. Engel et al., "Supersensitive Detection of Explosives by Silicon Nanowire Arrays," *Angewandte Chemie International Edition* 49 (2010): 6830–35, doi:10.1002/anie.201000847.
22. T. Jezierski et al., "Factors Affecting Drugs and Explosives Detection by Dogs in Experimental Tests," Third Canine Science Forum (Barcelona, July 25–27, 2012).
23. Goldblatt, Gazit, and Terkel, "Olfaction and Explosives Detector Dogs."
24. I. Jalakas, *Den nyttiga nosen* (Bilda Forlag, 2000).
25. GICHD, *Remote Explosive Scent Tracing, REST* (Geneva International Centre for Humanitarian Demining, 2011).
26. E. Lotspeich, K. Kitts, and J. Goodpaster, "Headspace Concentrations of Explosive Vapors in Containers Designed for Canine Testing and Training: Theory, Experiment, and Canine Trials," *Forensic Science International* 220 (2012): 130–34, doi:10.1016/j.forsciint.2012.02.009.
27. L. Lazarowski and D. C. Dorman, "Explosives Detection by Military Working Dogs: Olfactory Generalization from Components to Mixtures," *Applied Animal Behaviour Science* 151 (2014): 84–93, doi:http://dx.doi.org/10.1016/j.applanim.2013.11.010.
28. W. Kranz et al., "On the Smell of Composition C-4," *Forensic Science International* 236 (2014): 157–63, doi:http://dx.doi.org/10.1016/j.forsciint.2013.12.012.
29. Furton and Myers, "The Scientific Foundation and Efficacy of the Use of Canines as Chemical Detectors for Explosives"; M. Goldish, *Bomb-Sniffing Dogs* (Bearport Publishing, 2012).
30. Ibid.
31. "Vapor Wake Detections Dogs Bring New Levels of Security," *Auburn News*, http://www.alumni.auburn.edu/vapor-wake-detection-dogs-bring-new-levels-of-security/.
32. T. Harding, "Bomb Sniffer Dog in Line for Animal VC," February 27, 2006, http://www.telegraph.co.uk/news/uknews/1511557/Bomb-sniffer-dog-in-line-for-animal-VC.html.

33. Agence France-Presse, "Nosey Barker Finds Stash," April 3, 2003, http://www.smh.com.au/articles/2003/04/02/1048962819532.html.
34. W. S. Helton, ed., *Canine Ergonomics: The Science of Working Dogs* (CRC Press, 2009); Allsopp, *Cry Havoc*.
35. I. G. McLean, GICHD, *Designer Dogs: Improving the Quality of Mine Detection Dogs* (The Geneva International Centre for Humanitarian Demining, 2001), https://www.files.ethz.ch/isn/26650/Catalogue_Designer_Dogs.pdf; I. G. McLean, GICHD, *Mine Detection Dogs: Training, Operations and Odour Detection* (The Geneva International Centre for Humanitarian Demining, 2003), http://www.gichd.org/fileadmin/GICHD-resources/rec-documents/MDD.pdf.
36. L. Davner, "Decay! Dogs Hunting Damaged Wood," *Skogen* 5 (1986): 48.
37. M. Weisbord and K. Kachanoff, *Dogs with Jobs* (Pocket Books, 2000).
38. N. Hadzimujagic, "Effect of Inadequate Care for Mine Detection Dogs on Performance Quality in the Mine Action Operations," e-mail report, Mine Detection Dog Center in Bosnia and Herzegovina (2013).
39. GICHD, *Remote Explosive Scent Tracing, REST*.
40. Geneva International Centre for Humanitarian Demining, http://www.gichd.org/.
41. T. Berntsen, PowerPoint presentation, sent per e-mail, 2013.
42. Norsk Folkehjelp, http://www.folkehjelp.no/Vaart-arbeid/Miner-og-vaapen/Hva-vi-gjoer/Mine-og-eksplosivrydding/Minehunder.
43. Ibid.
44. A. Goth, I. G. McLean, and J. Trevelyan, "How Do Dogs Detect Landmines? A Summary of Research Results," in *Odour Detection: The Theory and the Practice* (Geneva International Centre for Humanitarian Demining, 2002), 195–208, http://www.gichd.org/fileadmin/pdf/publications/MDD/MDD_ch5_part1.pdf.
45. McLean, *Designer Dogs*; McLean, *Mine Detection Dogs*.
46. J. M. Phelan and J. L. Barnett, "Chemical Sensing Thresholds for Mine Detection Dogs," *Proceedings Of SPIE* 4742 (2002): 532, http://dx.doi.org/10.1117/12.479126.
47. I. G. McLean and R. J. Sargisson, "Optimising the Use of REST for Mine Detection," *Journal of Mine Action* 8 (2004): 100–104.
48. Ibid.
49. B. M. Jones, "Applied Behavior Analysis Is Ideal for the Development of a Land Mine Detection Technology Using Animals," *Behavior Analyst* 34 (2011): 55–73.
50. R. Fjellanger, E. K. Andersen, and I. G. McLean, "A Training Program for Filter-Search Mine Detection Dogs," *International Journal of Comparative Psychology* 15 (2002): 277–86.
51. McLean and Sargisson, "Optimising the Use of REST for Mine Detection."
52. Fjellanger, Andersen, and McLean, "A Training Program for Filter-Search Mine Detection Dogs."
53. McLean, *Mine Detection Dogs*.
54. Fjellanger, Andersen, and McLean, "A Training Program for Filter-Search Mine Detection Dogs."

55. V. Joynt, "The Mechem Explosive and Drug Detection System (MEDDS)," in *Mine Detection Dogs: Training, Operations and Odour Detection* (The Geneva International Centre for Humanitarian Demining, 2003), 165–74, http://www.gichd.org/fileadmin/GICHD-resources/rec-documents/MDD.pdf.
56. McLean and Sargisson, "Optimising the Use of REST for Mine Detection."
57. STU-100, "Scent Transfer Unit," http://www.stu100.com/; A. Uddqvist and I. Roberthson, *Improvement of Sampling System for Remote Explosive Scent Tracing* (2010).
58. GICHD, *Remote Explosive Scent Tracing, REST*.

CHAPTER TEN

1. L. Fisher, "The One Pupil Allowed to Sniff in Class: How Shirley the Labrador Uses Her Nose to Stop Rebecca, 7, Falling into a Coma," *Daily Mail Online* February 8, 2011.
2. K. M. C. Cline, "Psychological Effects of Dog Ownership: Role Strain, Role Enhancement, and Depression," *Journal of Social Psychology* 150 (2010): 117–31; D. A. Marcus, "The Science Behind Animal-Assisted Therapy," *Current Pain and Headache Reports* 17 (2013), doi:10.1007/s11916-013-0322-2; D. L. Wells, "Domestic Dogs and Human Health: An Overview," *British Journal of Health Psychology* 12 (2007): 145–56, doi:10.1348/135910706x103284.
3. A. T. B. Edney, "Companion Animals and Human Health," *Veterinary Record* 130 (1992): 285–87.
4. Cline, "Psychological Effects of Dog Ownership."
5. D. L. Wells, "Dogs as a Diagnostic Tool for Ill Health in Humans," *Alternative Therapies in Health and Medicine* 18 (2012): 12–17; L. R. Bijland, M. K. Bomers, and Y. M. Smulders, "Smelling the Diagnosis: A Review on the Use of Scent in Diagnosing Disease," *Netherlands Journal of Medicine* 71 (2013): 300–307.
6. M. Smith, "The Use of Smell in Differential Diagnosis," *Lancet* 25 (1982): 1452–53; C. E. Garner et al., "Volatile Organic Compounds from Feces and Their Potential for Diagnosis of Gastrointestinal Disease," *FASEB Journal* 21 (2007): 1675–88.
7. R. E. Brown, "What Is the Role of the Immune System in Determining Individually Distinct Body Odors?," *International Journal of Immunopharmacology* 17 (1995): 655–61, doi:10.1016/0192-0561(95)00052-4.
8. T. Tuuminen, "Urine as a Specimen to Diagnose Infections in Twenty-First Century: Focus on Analytical Accuracy," *Frontiers in Immunology* 3 (2012): 1–6.
9. Smith, "The Use of Smell in Differential Diagnosis."
10. O. Fyrand, *Det gåtefulle språket. Om hudens kommunikasjon* (Universitetesforlaget AS, 1996).
11. J. McSherry, "Sniffing Out the Diagnosis," *Canadian Medical Association Journal* 135 (1986): 1070.
12. Fyrand, *Det gåtefulle språket*; K. Smith, G. F. Thompson, and H. D. Koster, "Sweat in Schizophrenic Patients: Identification of Odorous Substance," *Science* 166 (1969): 398–99, doi:10.1126/science.166.3903.398.

13. G. S. Settles, "Sniffers: Fluid-Dynamic Sampling for Olfactory Trace Detection in Nature and Homeland Security: The 2004 Freeman Scholar Lecture," *Journal of Fluids Engineering, Transactions of the ASME* 127 (2005): 189–218, doi:10.1115/1.1891146.
14. International Diabetes Foundation, IDF Diabetes Atlas, 7th ed., (IDF, 2015), https://www.idf.org/our-activities/advocacy-awareness/resources-and-tools/13:diabetes-atlas-seventh-edition.html.
15. Diabetes Forbundet, www.diabetes.no.
16. C. N. Tassopoulos, D. Barnett, and T. R. Fraser, "Breath-Acetone and Blood-Sugar Measurements in Diabetes," *Lancet* 293 (1969): 1282–86; M. Fleischer et al., "Detection of Volatile Compounds Correlated to Human Diseases through Breath Analysis with Chemical Sensors," *Sensors and Actuators B-Chemical* 83 (2002): 245–49, doi:10.1016/s0925-4005(01)01056-5.
17. Diabetes Forbundet, www.diabetes.no.
18. Ibid.
19. M. B. O'Connor, C. O'Connor, and C. H. Walsh, "A Dog's Detection of Low Blood Sugar: A Case Report," *Irish Journal of Medical Science* 177 (2008): 155–57, doi:10.1007/s11845-008-0128-0.
20. M. Chen et al., "Non-Invasive Detection of Hypoglycaemia Using a Novel, Fully Biocompatible and Patient Friendly Alarm System," *BMJ* 321 (2000): 1565–66; D. L. Wells, S. W. Lawson, and A. N. Siriwardena, "Canine Responses to Hypoglycemia in Patients with Type 1 Diabetes," *Journal of Alternative and Complementary Medicine* 14 (2008): 1235–41, doi:10.1089/acm.2008.0288.
21. K. Dehlinger et al., "Can Trained Dogs Detect a Hypoglycemic Scent in Patients with Type 1 Diabetes?" *Diabetes Care* 36 (2013): E98–E99, doi:10.2337/dc12-2342.
22. I. Tauveron et al., "Canine Detection of Hypoglycaemic Episodes Whilst Driving," *Diabetic Medicine* 23 (2006): 335, doi:10.1111/j.1464-5491.2006.01820.x.
23. V. McAulay, I. J. Deary, and B. M. Frier, "Symptoms of Hypoglycaemia in People with Diabetes," *Diabetic Medicine* 18 (2001): 690–705, doi:10.1046/j.1464--5491.2001.00620.x.
24. O'Connor, O'Connor, and Walsh, "A Dog's Detection of Low Blood Sugar."
25. Tauveron et al., "Canine Detection of Hypoglycaemic Episodes Whilst Driving."
26. Wells, Lawson, and Siriwardena, "Canine Responses to Hypoglycemia in Patients with Type 1 Diabetes."
27. N. Rooney, S. Morant, and C. Guest, "Investigation into the Value of Trained Glycaemia Alert Dogs to Clients with Type 1 Diabetes," *PLOS ONE* 8 (2013): e69921.
28. "Dog Chews Off Owner's Toe and May Have Saved His Life," NBCNews.com, August 3, 2010, http://bodyodd.nbcnews.com/_news/2010/08/03/4807400-dog-chews-off-owners-toe-and-may-have-saved-his-life?lite.
29. A. Johansen, e-mail message to author, August 8, 2013.
30. V. Zimmerman, *Dog, a Diabetic's Best Friend Training Guide: Train Your Own Diabetic and Glycemic Alert Dog* (Black & White Edition, 2011); R. Martinez and

S. M. Barns, *Training Your Diabetic Alert Dog* (self-published, 2013); N. Bondarenko, "Correlation between Assessment, Selection and Training Outcomes of a Puppy to Be Used as a Diabetic Alert Dog," Third Canine Science Forum (Barcelona, July 25–27, 2012).

31. A. Johansen, e-mail message to author, August 8, 2013; T.-W. Skille, "En ekte blodhund," *Diabetes* 3 (2013): 34–35.
32. A. Jemal et al., "Global Cancer Statistics," *CA: Cancer Journal for Clinicians* 61 (2011): 69–90.
33. Kreftregisteret, http://www.kreftregisteret.no/.
34. H. Williams and A. Pembroke, "Sniffer Dogs in the Melanoma Clinic," *Lancet* 1 (1989): 734.
35. J. Church and H. Williams, "Another Sniffer Dog for the Clinic?" *Lancet* 358 (2001): 930, doi:10.1016/s0140-6736(01)06065-2.
36. J. S. Welsh, D. Barton, and H. Ahuja, "A Case of Breast Cancer Detected by a Pet Dog," *Community Oncology* (July/August 2005): 325–26.
37. Wells, "Dogs as a Diagnostic Tool for Ill Health in Humans."
38. S. C. Balsetro and H. R. Correia, "Is Olfactory Detection of Human Cancer by Dogs Based on Major Histocompatibility Complexdependent Odour Components? A Possible Cure and a Precocious Diagnosis of Cancer," *Medical Hypotheses* 66 (2006): 270–72, doi:10.1016/j.mehy.2005.08.027.
39. C. M. Willis et al., "Olfactory Detection of Human Bladder Cancer by Dogs: Proof of Principle Study," *British Medical Journal* 329 (2004): 712–14A, doi:10.1136/bmj.329.7468.712; D. Pickel et al., "Evidence for Canine Olfactory Detection of Melanoma," *Applied Animal Behaviour Science* 89 (2004): 107–16, doi:10.1016/j.applanim.2004.04.008; T. Jezierski, M. Walczak, and A. Gorecka, "Canine Olfactory Detection of Human Cancer Odor Markers," *Polish Academy of Sciences, Annual Report* (2008), 64–65; M. Walczak et al., "Impact of Individual Training Parameters and Manner of Taking Breath Odor Samples on the Reliability of Canines as Cancer Screeners," *Journal of Veterinary Behavior: Clinical Applications and Research* 7 (2012): 283–94, doi:10.1016/j.jveb.2012.01.001; M. McCulloch et al., "Diagnostic Accuracy of Canine Scent Detection in Early- and Late-Stage Lung and Breast Cancers," *Integrative Cancer Therapies* 5 (2006): 30–39, doi:10.1177/1534735405285096; B. Buszewski et al., "Identification of Volatile Lung Cancer Markers by Gas Chromatography-Mass Spectrometry: Comparison with Discrimination by Canines," *Analytical and Bioanalytical Chemistry* 404 (2012): 141–46, doi:10.1007/s00216-012-6102-8; R. Ehmann et al., "Canine Scent Detection in the Diagnosis of Lung Cancer: Revisiting a Puzzling Phenomenon," *European Respiratory Journal* 39 (2012): 669–76, doi:10.1183/09031936.00051711; T. Amundsen et al., "Can Dogs Smell Lung Cancer? First Study Using Exhaled Breath and Urine Screening in Unselected Patients with Suspected Lung Cancer," *Acta oncologica* (Stockholm, Sweden) 53 (2014): 307–15, doi:10.3109/0284186x.2013.819996; Jezierski, Walczak, and Gorecka, "Canine Olfactory Detection

of Human Cancer Odor Markers"; Walczak et al., "Impact of Individual Training Parameters and Manner of Taking Breath Odor Samples on the Reliability of Canines as Cancer Screeners"; McCulloch et al., "Diagnostic Accuracy of Canine Scent Detection in Early- and Late-Stage Lung and Breast Cancers"; Buszewski et al., "Identification of Volatile Lung Cancer Markers by Gas Chromatography-Mass Spectrometry"; G. Horvath et al., "Human Ovarian Carcinomas Detected by Specific Odor," *Integrative Cancer Therapies* 7 (2008): 76–80, doi:10.1177/1534735408319058; G. Horvath, H. Andersson, and S. Nemes, "Cancer Odor in the Blood of Ovarian Cancer Patients: A Retrospective Study of Detection by Dogs during Treatment, 3 and 6 Months Afterwards," *BMC Cancer* 13 (2013): 396–401; R. T. Gordon et al., "The Use of Canines in the Detection of Human Cancers," *Journal of Alternative and Complementary Medicine* 14 (2008): 61–67, doi:10.1089/acm.2006.6408; H. Sonoda et al., "Colorectal Cancer Screening with Odour Material by Canine Scent Detection," *Gut* 60 (2011): 814–19, doi:10.1136/gut.2010.218305.

40. Willis et al., "Olfactory Detection of Human Bladder Cancer by Dogs."
41. P. Spanel et al., "Analysis of Formaldehyde in the Headspace of Urine from Bladder and Prostate Cancer Patients Using Selected Ion Flow Tube Mass Spectrometry," *Rapid Communications Mass Spectrometry* 13 (1999): 1354–59, doi:10.1002/(sici)1097-0231(19990730)13:14<1354::aid-rcm641>3.0.co;2-j.
42. Willis et al., "Olfactory Detection of Human Bladder Cancer by Dogs."
43. K. Wakamatsu et al., "Evaluation of 5-S-Cysteinyidopa as a Marker of Melanoma Progression: 10 years' Experience," *Melanoma Research* 12 (2002): 245–53, doi:10.1097/00008390-200206000-00008.
44. Pickel et al., "Evidence for Canine Olfactory Detection of Melanoma."
45. McCulloch et al., "Diagnostic Accuracy of Canine Scent Detection in Early- and Late-Stage Lung and Breast Cancers."
46. G. Horvath, J. Chilo, and T. Lindblad, "Different Volatile Signals Emitted by Human Ovarian Carcinoma and Healthy Tissue," *Future Oncology* 6 (2010): 1043–49, doi:10.2217/fon.10.60.
47. Horvath et al., "Human Ovarian Carcinomas Detected by Specific Odor."
48. G. Horvath, H. Andersson, and G. Paulsson, "Characteristic Odour in the Blood Reveals Ovarian Carcinomam," *BMC Cancer* 10 (2010), doi:64310.1186/1471-2407-10-643.
49. Sonoda et al., "Colorectal Cancer Screening with Odour Material by Canine Scent Detection."
50. J. N. Cornu et al., "Olfactory Detection of Prostate Cancer by Dogs Sniffing Urine: A Step Forward in Early Diagnosis," *European Urology* 59 (2011): 197–201, doi:10.1016/j.eururo.2010.10.006.
51. Wells, "Dogs as a Diagnostic Tool for Ill Health in Humans."
52. Walczak et al., "Impact of Individual Training Parameters and Manner of Taking Breath Odor Samples on the Reliability of Canines as Cancer Screeners."

53. E. Moser and M. McCulloch, "Canine Scent Detection of Human Cancers: A Review of Methods and Accuracy," *Journal of Veterinary Behavior: Clinical Applications and Research* 5 (2010): 145–52, doi:10.1016/j.jveb.2010.01.002.
54. G. Horvath, e-mail message to author, September 21, 2013.
55. Walczak et al., "Impact of Individual Training Parameters and Manner of Taking Breath Odor Samples on the Reliability of Canines as Cancer Screeners."
56. U. Ellervik, *Ond kemi. Berättelser om människor, mord och molekyler* (Fri tanke forlag, 2011).
57. G. Lippi and G. Cervellin, "Canine Olfactory Detection of Cancer versus Laboratory Testing: Myth or Opportunity?" *Clinical Chemistry and Laboratory Medicine* 50 (2012): 435–39, doi:10.1515/cclm.2011.672.
58. A. S. Bjartell, "Dogs Sniffing Urine: A Future Diagnostic Tool or a Way to Identify New Prostate Cancer Markers?" *European Urology* 59 (2011): 202–3, doi:10.1016/j.eururo.2010.10.033.
59. F. Rock, N. Barsan, and U. Weimar, "Electronic Nose: Current Status and Future Trends," *Chemical Reviews* 108 (2008): 705–25.
60. Bijland, Bomers, and Smulders, "Smelling the Diagnosis."
61. S. H. Svendsen, "Vil lære hunden å lukte eggstokkreft," *VG Nett*, September 10, 2013.
62. Horvath, Andersson, and Nemes, "Cancer Odor in the Blood of Ovarian Cancer Patients."
63. Jezierski, Walczak, and Gorecka, "Canine Olfactory Detection of Human Cancer Odor Markers."
64. E. P. McCarthy et al., "Mammography Use Helps to Explain Differences in Breast Cancer Stage at Diagnosis between Older Black and White Women," *Annals of Internal Medicine* 128 (1998): 729–36.
65. A. Ulanowska et al., "Hyphenated and Unconventional Methods for Searching Volatile Cancer Biomarkers," *Ecological Chemistry and Engineering S/Chemia I Inzynieria Ekologiczna S* 17 (2010): 9–23.
66. W. S. Helton, ed., *Canine Ergonomics: The Science of Working Dogs* (CRC Press, 2009).
67. M. Leja, H. Liu, and H. Haick, "Breath Testing: The Future for Digestive Cancer Detection," *Expert Review of Gastroenterology and Hepatology* 75 (2013): 389–91; D. F. Altomare, "Exhaled Volatile Organic Compounds Identify Patients with Colorectal Cancer," *British Journal of Surgery* 100 (2013): 144–50, doi:10.1002/bjs.9149.
68. Amundsen et al., "Can Dogs Smell Lung Cancer?"
69. S. O. Isaksen, "Loddpenger til krefthund," *Røyken og Hurums avis*, October 19, 2012.
70. D. A. Marcus, "The Science Behind Animal-Assisted Therapy," *Current Pain and Headache Reports* 17 (2013), doi:10.1007/s11916-013-0322-2; A. Edney, "Dogs and Human Epilepsy," *Veterinary Record* 132 (1993): 337–38; V. Strong, S. W. Brown, and R. Walker, "Seizure-Alert Dogs—Fact or Fiction?" *Seizure: European Journal of Epilepsy* 8 (1999): 62–65, doi:10.1053/seiz.1998.0250.
71. Edney, "Dogs and Human Epilepsy."

72. Peanut Detector Dogs, http://www.peanutdogs.com/; B. Newsome, "Peanut-Sniffing Dog Is Allergic Girl's Best Friend," *Gazette*, February 16, 2009.
73. C. Holst, "Meghan kan dø af at spise en nød: min hund reder meg," *Ude & hjemme* 34 (2013): 4–5.
74. EENP: Eyes, Ears, Nose & Paws, http://eenp.org/main/KKandJJ.
75. M. K. Bomers et al., "Using a Dog's Superior Olfactory Sensitivity to Identify *Clostridium difficile* in stools and Patients: Proof of Principle Study," *British Medical Journal* 345 (2012), doi:10.1136/bmj.e7396.
76. M. Syhre and S. T. Chambers, "The Scent of *Mycobacterium tuberculosis*," *Tuberculosis* 88 (2008): 317–23, doi:10.1016/j.tube.2008.01.002.
77. D. M. Suckling and R. L. Sagar, "Honeybees *Apis mellifera* Can Detect the Scent of *Mycobacterium tuberculosis*," *Tuberculosis* 91 (2011): 327–28, doi:10.1016/j.tube.2011.04.008; G. F. Mgode et al., "*Mycobacterium tuberculosis* Volatiles for Diagnosis of Tuberculosis by Cricetomys Rats," *Tuberculosis* 92 (2012): 535–42, doi:10.1016/j.tube.2012.07.006; G. F. Mgode et al., "Ability of Cricetomys Rats to Detect *Mycobacterium tuberculosis* and Discriminate It from Other Microorganisms," *Tuberculosis* 92 (2012): 182–86, doi:10.1016/j.tube.2011.11.008; A. Poling et al., "Tuberculosis Detection by Giant African Pouched Rats," *Behavior Analyst* 34 (2011): 47–54.
78. J. Tasker, "Sniffer Dogs to Detect Bovine TB?" *Farmers Weekly* October 1, 2012.
79. Tuuminen, "Urine as a Specimen to Diagnose Infections in Twenty-First Century"; Church and Williams, "Another Sniffer Dog for the Clinic?"
80. You can be kept up-to-date at the Medical Detection Dogs home page: http://medicaldeteciondogs.org.uk/.
81. J. W. Simons, "How My Beloved Dog Found My Cancer," *Telegraph*, March 17, 2013, http://www.telegraph.co.uk/lifestyle/pets/9935073/How-my-beloved-dog-found-my-cancer.html.
82. J. Potts, "The Dogs Who Can Smell Cancer," *Telegraph Blogs*, November 13, 2012.
83. Medical Detection Dogs, http://medicaldetectiondogs.org.uk/; C. Guest, e-mail message to author, June 6, 2014.

CHAPTER ELEVEN

1. D. J. Shaver, "Analysis of Kemp's Ridley Imprinting and Headstart Project at Padra Island National Seashore, Texas, 1978–88, with Subsequent Nesting and Stranding Records on the Texas Coast," *Chelonian Conservation and Biology* 44 (2005): 846–59.
2. "Ranger Ridley to the Rescue: A Texas Cairn Joins the Fight to Save a Rare Sea Turtle," *The Cairn Terrier Club of America* (Spring 2013): 63–64.
3. D. A. S. Woollett, A. Hurt, and N. L. Richards, "The Current and Future Roles of Free-Ranging Detection Dogs in Conservation Efforts," in *Free-Ranging Dogs and Wildlife Conservation*, ed. M. E. Gompper (Oxford University Press, 2014), 239–64.

4. D. K. Dahlgren et al., "Use of Dogs in Wildlife Research and Management," in *The Wildlife Techniques Manual*, vol. 1, ed. N. J. Silvy, 7th ed. (Johns Hopkins University Press, 2012), 140–53.
5. K. Campbell and C. J. Donlan, "Feral Goat Eradications on Islands," *Conservation Biology* 19 (2005): 1362–74, doi:10.1111/j.1523-1739.2005.00228.x; C. Browne, K. Stafford, and R. Fordham, "The Use of Scent-Detection Dogs," *Irish Veterinary Journal* 59 (2006): 97–104; P. E. Cowan, "The Eradication of Introduced Australian Brushtail Possums, *Trichosurus vulpecula*, from Kapiti Island, a New Zealand Nature Reserve," *Biological Conservation* 61 (1992): 217–26, doi:10.1016/0006-3207(92)91119-d; A. Hurt and D. A. Smith, "Conservation Dogs," in *Canine Ergonomics: The Science of Working Dogs*, ed. W. S. Helton (CRC Press, 2009), 175–94.
6. D. A. Smith et al., "Detection and Accuracy Rates of Dogs Trained to Find Scats of San Joaquin Kit Foxes (*Vulpes macrotis mutica*)," *Animal Conservation* 6 (2003): 339–46, doi:10.1017/s136794300300341x.
7. Hurt and Smith, "Conservation Dogs."
8. Smith et al., "Detection and Accuracy Rates of Dogs Trained to Find Scats of San Joaquin Kit Foxes (*Vulpes macrotis mutica*)."
9. R. A. Long et al., "Comparing Scat Detection Dogs, Cameras, and Hair Snares for Surveying Carnivores," *Journal of Wildlife Management* 71 (2007): 2018–25, doi:10.2193/2006-292; D. A. Smith et al., "Canine Assistants for Conservationists," *Science* 291 (2001): 435; C. Vynne et al., "Effectiveness of Scat-Detection Dogs in Determining Species Presence in a Tropical Savanna Landscape," *Conservation Biology* 25 (2011): 154–62, doi:10.1111/j.1523-1739.2010.01581.x.
10. Smith et al., "Canine Assistants for Conservationists."
11. H. A. Robertson and J. R. Fraser, "Use of Trained Dogs to Determine the Age Structure and Conservation Status of Kiwi *Apteryx* spp. Populations," *Bird Conservation International* 19 (2009): 121–29, doi:10.1017/s0959270908007673; R. Colbourne, "Little Spotted Kiwi (*Apteryx owenii*): Recruitment and Behavior of Juveniles on Kapiti Island, New Zealand, *Journal of the Royal Society of New Zealand* 22 (1992): 321–28.
12. S. Hill and J. Hill, *Richard Henry of Resolution Island* (John McIndoe Ltd., 1987), 364.
13. C. M. Browne, "The Use of Dogs to Detect New Zealand Reptile Scents" (master's thesis, Massey University, 2005).
14. A. Gsell et al., "The Success of Using Trained Dogs to Locate Sparse Rodents in Pest-Free Sanctuaries," *Wildlife Research* 37 (2010): 39–46, doi:10.1071/wr09117; I. Shapira, F. Buchanan, and D. H. Brunton, "Detection of Caged and Free-Ranging Norway Rats *Rattus norvegicus* by a Rodent Sniffing Dog on Browns Island, Auckland, New Zealand," *Conservation Evidence* 8 (2011): 38–42.
15. K. Vincent, e-mail message to author, April 17, 2013.
16. W. R. Skinner, D. P. Snow, and N. F. Payne, "A Capture Technique for Juvenile Willow Ptarmigan," *Wildlife Society Bulletin* 26 (1998): 111–12; H. C. Pedersen et al., "Weak Compensation of Harvest Despite Strong Density-Dependent Growth in

Willow Ptarmigan," *Proceedings of the Royal Society B: Biological Sciences* 271 (2004): 381–85, doi:10.1098/rspb.2003.2599; B. K. Sandercock et al., "Is Hunting Mortality Additive or Compensatory to Natural Mortality? Effects of Experimental Harvest on the Survival and Cause-Specific Mortality of Willow Ptarmigan," *Journal of Animal Ecology* 80 (2011): 244–58, doi:10.1111/j.1365-2656.2010.01769.x.

17. Campagna et al., "Potential Semiochemical Molecules from Birds: A Practical and Comprehensive Compilation of the Last 20 Years Studies," *Chemical Senses* 37 (2012): 3–25, doi:10.1093/chemse/bjr067.
18. J. Reneerkens, T. Piersma, and J. S. S. Damste, "Switch to Diester Preen Waxes May Reduce Avian Nest Predation by Mammalian Predators Using Olfactory Cues," *Journal of Experimental Biology* 208 (2005): 4199–202, doi:10.1242/jeb.01872.
19. I. Espelien, *Lundehundboka* (Vigmostad & Bjørke, 2012).
20. http://www.nrk.no/nordnytt/lundehunder-sikrer-flyplassen-1.11099894; M. Evenseth, e-mail message to author, August 2013; Norsk Lundehund Klubb, "Norsk Lundehund," *Hundesport* 2 (2010): 39–41; M. Evenseth, e-mail message to author, June 13, 2014.
21. L. M. Klauber, *Rattlesnakes: Their Habits, Life Histories, and Influence on Mankind* (University of California Press, 1956).
22. D. J. Stevenson et al., "Using a Wildlife Detector Dog for Locating Eastern Indigo Snakes (*Drymarchon couperi*)," *Herpetological Review* 41 (2010): 437–42.
23. I. Gazit and J. Terkel, "Domination of Olfaction over Vision in Explosives Detection by Dogs," *Applied Animal Behaviour Science* 82 (2003): 65–73, doi:10.1016/s0168-1591(03)00051-0; D. A. Smith et al., "Detection and Accuracy Rates of Dogs Trained to Find Scats of San Joaquin Kit Foxes (*Vulpes macrotis mutica*)."
24. D. Grandjean and F. Haymann, *Hunde-encyklopedi* (RK Grafisk, 2010).
25. M. E. Dorcas et al., "Severe Mammal Declines Coincide with Proliferation of Invasive Burmese Pythons in Everglades National Park," *Proceedings of the National Academy of Sciences of the United States of America* 109 (2012): 2418–22, doi:10.1073/pnas.1115226109.
26. C. Martin, "A Giant Battle: Auburn Canines Help in Search for Everglades' Pythons," Auburn University, February 15, 2012, http://ocm.auburn.edu/featured_story/pythons_dogs.html.
27. G. H. Rodda and T. H. Fritts, "The Impact of the Introduction of the Colubrid Snake *Boiga irregularis* on Guam Lizards," *Journal of Herpetology* 26 (1992): 166–74, doi:10.2307/1564858.
28. L. R. Rawlings et al., "Phylogenetic Analysis of the Brown Treesnake, *Boiga irregularis*, Particularly Relating to a Population on Guam," Brown Treesnake Research Symposium (Honolulu, Hawaii, 1998).
29. Rodda and Fritts, "The Impact of the Introduction of the Colubrid Snake *Boiga irregularis* on Guam Lizards."
30. T. H. Fritts and M. J. McCoid, "Predation by the Brown Treesnake (*Boiga irregularis*) on Poultry and Other Domesticated Animals on Guam," *Snake* 23 (1991): 75–80.

31. T. H. Fritts, M. J. McCoid, and R. L. Haddock, "Risks to Infants on Guam from Bites of the Brown Treesnake (*Boiga irregularis*)," *American Journal of Tropical Medicine and Hygiene* 42 (1990): 606–11; T. H. Fritts, M. J. McCoid, and R. L. Haddock, "Symptoms and Circumstances Associated with Bites by the Brown Treesnake (Colubridae: *Boiga irregularis*) on Guam," *Journal of Herpetology* 28 (1994): 27–33.
32. T. H. Fritts, N. J. Scott Jr., and J. A. Savidge, "Activity of the Arboreal Brown Tree Snake (*Boiga irregularis*) on Guam as Determined by Electrical Power Outages," *Snake* 19 (1987): 51–58.
33. D. S. Vice and M. E. Pitzler, "Brown Treesnake Conrol: Economy of Scales," (Colorado, National Wildlife Research Center, Proceeding of the Third NWRC Special Symposium, 2002), 127–31.
34. J. J. Johnston et al., "Ecotoxicological Risks of Potential Toxicants for Brown Tree Snake Control on Guam" (2001), http://digitalcommons.unl.edu/icwdm_usdanwrc/590/.
35. D. S. Vice and R. M. Engeman, "Brown Tree Snake Discoveries during Detector Dog Inspections Following Supertyphoon Paka," *Micronesica* 33 (2000): 105–10.
36. R. M. Engeman et al., "Sustained Evaluation of the Effectiveness of Detector Dogs for Locating Brown Tree Snakes in Cargo Outbound from Guam," *International Biodeterioration and Biodegradation* 49 (2002): 101–6, doi:10.1016/s0964-8305(01)00109-3.
37. D. S. Vice, e-mail message to author, April 2013.
38. J. A. Savidge et al., "Canine Detection of Free-Ranging Brown Treesnakes on Guam," *New Zealand Journal of Ecology* 35 (2011): 174–81.
39. M. T. Christy et al., "Modelling Detection Probabilities to Evaluate Management and Control Tools for an Invasive Species," *Journal of Applied Ecology* 47 (2010): 106–13, doi:10.1111/j.1365-2664.2009.01753.x.
40. J. Paula et al., "Dogs as a Tool to Improve Bird-Strike Mortality Estimates at Wind Farms," *Journal for Nature Conservation* 19 (2011): 202–8, doi:10.1016/j.jnc.2011.01.002.
41. S. M. J. G. Steyaert et al., "Male Reproductive Strategy Explains Spatiotemporal Segregation in Brown Bears," *Journal of Animal Ecology* 82 (2013): 836–45, doi:10.1111/1365-2656.12055; S. M. J. G. Steyaert et al., "Infanticide as a Male Reproductive Strategy Has a Nutritive Risk Effect in Brown Bears," *Biology Letters* (2013).
42. M. Clapham et al., "A Hypothetico-Deductive Approach to Assessing the Social Function of Chemical Signalling in a Non-Territorial Solitary Carnivore," *PLOS ONE* 7 (2012), doi:10.1371/journal.pone.0035404.
43. I. Hansen, "Kadaversøkende hunder," *Bioforsk tema* 6 (2011): 1–6.
44. Ibid.
45. I. Hansen and L. J. Hind, "Erfaringer med bruk av kadaversøkende hunder i Norge," *Bioforsk Report* 4, no. 130 (2009).

46. H. J. Homan, G. Linz, and B. D. Peer, "Dogs Increase Recovery of Passerine Carcasses in Dense Vegetation," *Wildlife Society Bulletin* 29 (2001): 292–96.
47. K. Bevanger et al., "Optimal Design and Routing of Power Lines: Ecological, Technical and Economic Perspectives (OPTIPOL)" (Progress report 2010. NINA rapport 619, 2009), 51; K. Bevanger and H. Brøseth, "Bird Collisions with Power Lines: An Experiment with Ptarmigan (*Lagopus* spp.)," *Biological Conservation* 99 (2001): 341–46, doi:10.1016/s0006-3207(00)00217-2; K. Bevanger and H. Brøseth, "Impact of Power Lines on Bird Mortality in a Subalpine Area," *Animal Biodiversity and Conservation* 27 (2004): 67–77.
48. O. Reitan, "Søk etter døde fugler i Smøla vindpark 2011-årsrapport," *NINA Rapport* 790 (2012); O. Reitan, "Søk etter døde fugler i Smøla vindpark 2012-årsrapport," *NINA Rapport* 925 (2013).
49. E. B. Arnett, "A Preliminary Evaluation on the Use of Dogs to Recover Bat Fatalities at Wind Energy Facilities," *Wildlife Society Bulletin* 34 (2006): 1440–45, doi: 10.2193/0091-7648(2006)34[1440:apeotu]2.0.co;2.
50. F. Mathews et al., "Effectiveness of Search Dogs Compared with Human Observers in Locating Bat Carcasses at Wind-Turbine Sites: A Blinded Randomized Trial," *Wildlife Society Bulletin* 37 (2013): 34–40.
51. S. K. Wasser et al., "Using Detection Dogs to Conduct Simultaneous Surveys of Northern Spotted (*Strix occidentalis caurina*) and Barred Owls (*Strix varia*)," *PLOS ONE* 7 (2012), doi:e42892 10.1371/journal.pone.0042892.
52. Smith et al., "Canine Assistants for Conservationists."
53. M. Parker, "Wildlife Detection Dogs: Specially Trained, These Canines Work Hard for Conservation," *Wildlife Professional* (2011): 47–49.
54. Smith et al., "Canine Assistants for Conservationists."
55. Long et al., "Comparing Scat Detection Dogs, Cameras, and Hair Snares for Surveying Carnivores."
56. R. A. Long et al., "Effectiveness of Scat Detection Dogs for Detecting Forest Carnivores," *Journal of Wildlife Management* 71 (2007): 2007–17, doi:10.2193/2006-230.
57. R. M. Rolland et al., "Faecal Sampling Using Detection Dogs to Study Reproduction and Health in North Atlantic Right Whales (*Eubalaena glacialis*)," *Journal of Cetacean Research and Management* 8 (2006): 121–25.
58. K. L. Ayres et al., "Distinguishing the Impacts of Inadequate Prey and Vessel Traffic on an Endangered Killer Whale (*Orcinus orca*) Population," *PLOS ONE* 7 (2012), doi:e36842 10.1371/journal.pone.0036842.
59. Ibid.
60. Long et al., "Comparing Scat Detection Dogs, Cameras, and Hair Snares for Surveying Carnivores."
61. Long et al., "Effectiveness of Scat Detection Dogs for Detecting Forest Carnivores."
62. Long et al., "Comparing Scat Detection Dogs, Cameras, and Hair Snares for Surveying Carnivores."

63. L. L. Kerley and G. P. Salkina, Using Scent-Matching Dogs to Identify Individual Amur Tigers from Scats," *Journal of Wildlife Management* 71 (2007): 1349–56, doi:10.2193/2006-361.
64. Parker, "Wildlife Detection Dogs."
65. D. A. Smith et al., "Assessment of Scat-Detection Dog Surveys to Determine Kit Fox Distribution," *Wildlife Society Bulletin* 33 (2005): 897–904, doi:10.2193/0091-7648(2005)33[897:aosdst]2.0.co;2; D. A. Smith et al., "Relative Abundance of Endangered San Joaquin Kit Foxes (*Vulpes macrotis mutica*) Based on Scat-Detection Dog Surveys," *Southwestern Naturalist* 51 (2006): 210–19, doi:10.1894/0038-4909(2006)51[210:raoesj]2.0.co;2; A. Hurt, B. Davenport, and E. Greene, "Training Dogs to Distinguish between Black Bear (*Ursus americanus*) and Grizzly Bear (*Ursus arctos*) Feces" (University of Montana Undergraduate Biology Journal, 2000).
66. Smith et al., "Detection and Accuracy Rates of Dogs Trained to Find Scats of San Joaquin Kit Foxes (*Vulpes macrotis mutica*)."
67. Smith et al., "Assessment of Scat-Detection Dog Surveys to Determine Kit Fox Distribution."
68. K. Ralls et al., "Changes in Kit Fox Defecation Patterns during the Reproductive Season: Implications for Noninvasive Surveys," *Journal of Wildlife Management* 74 (2010): 1457–62, doi:10.2193/2009-401.
69. K. Kauhala and L. Salonen, "Does a Non-Invasive Method—Latrine Surveys—Reveal Habitat Preferences of Raccoon Dogs and Badgers?" *Mammalian Biology* 77 (2012): 264–70, doi:10.1016/j.mambio.2012.02.007.
70. C. Vynne et al., "Effectiveness of Scat-Detection Dogs in Determining Species Presence in a Tropical Savanna Landscape."
71. Scandinavian Brown Bear Research Project, http://bearproject.info/.
72. F. Rosell, "Brown Bears Possess Anal Sacs and Secretions May Code for Sex," *Journal of Zoology* 283 (2011): 143–52, doi:10.1111/j.1469-7998.2010.00754.x.
73. S. M. Jojola et al., "Subadult Brown Bears (*Ursus arctos*) Discriminate between Unfamiliar Adult Male and Female Anal Gland Secretion," *Mammalian Biology* 77 (2012): 363–68, doi:10.1016/j.mambio.2012.05.003.
74. Clapham et al., "A Hypothetico-Deductive Approach to Assessing the Social Function of Chemical Signalling in a Non-Territorial Solitary Carnivore."
75. K. Ingdal and B. Lassen, "Chemical Communication in Brown Bears and Yellow-Bellied Marmots" (Master's thesis, Telemark University College, Bø i Telemark, Norway, 2008).
76. Clapham et al., "A Hypothetico-Deductive Approach to Assessing the Social Function of Chemical Signalling in a Non-Territorial Solitary Carnivore"; M. Clapham et al., "The Function of Strategic Tree Selectivity in the Chemical Signalling of Brown Bears," *Animal Behaviour* 85 (2013): 1351–57, doi:10.1016/j.anbehav.2013.03.026; M. Clapham et al., "Scent Marking in Wild Brown Bears: Time Investment, Motor Patterns and Age-Related Development," *Animal Behaviour* 94 (2014): 107–16.

77. K. M. Goodwin et al., "Trained Dogs Outperform Human Surveyors in the Detection of Rare Spotted Knapweed (*Centaurea stoebe*)," *Invasive Plant Science Management* 3 (2010): 113–21, doi:10.1614/ipsm-d-09-00025.1; K. M. Goodwin, "A Novel Method to Detect Knapweed (*Centaurea biebersteinii* DC) Using Specially Trained Canines" (Bozeman, Montana, 2005).
78. Goodwin et al., "Trained Dogs Outperform Human Surveyors in the Detection of Rare Spotted Knapweed (*Centaurea stoebe*)."
79. Ibid.
80. Dahlgren et al., "Use of Dogs in Wildlife Research and Management"; Parker, "Wildlife Detection Dogs."
81. Parker, "Wildlife Detection Dogs"; D. Fosdick, "Poachers Steal Public Plants: Collectors Join Other Thieves in Endangering Beneficial Wildlife," *Columbian* (Vancouver, WA), June 5, 2002.
82. D. L. Knobel et al., "Dogs, Disease, and Wildlife," in *Free-Ranging Dogs and Wildlife Conservation*, ed. M. E. Gompper (Oxford University Press, 2014), 144–69.
83. M. A. Weston and T. Stankowich, "Dogs as Agents of Disturbance," in *Free-Ranging Dogs and Wildlife Conservation*, ed. M. E. Gompper (Oxford University Press, 2014), 94–116.
84. P. B. Banks and J. V. Bryant, Four-Legged Friend or Foe? Dog Walking Displaces Native Birds from Natural Areas," *Biology Letters* 3 (2007): 611–13, doi:10.1098/rsbl.2007.0374.
85. J. S. Heaton et al., "Comparison of Effects of Humans versus Wildlife-Detector Dogs," *Southwestern Naturalist* 53 (2008): 472–79, doi:10.1894/pas-03.1.
86. W. F. Andelt, "Relative Effectiveness of Guarding-Dog Breeds to Deter Predation on Domestic Sheep in Colorado," *Wildlife Society Bulletin* 27 (1999): 706–14; J. F. Kamler et al., "Feral Dogs, *Canis familiaris*, Kill Coyote, *Canis latrans*," *Canadian Field-Naturalist* 117 (2003): 123–24.
87. J. Waters et al., "Testing a Detection Dog to Locate Bumblebee Colonies and Estimate Nest Density," *Apidologie* 42 (2011): 200–205, doi:10.1051/apido/2010056; S. O'Connor, K. J. Park, and D. Goulson, "Humans versus Dogs: A Comparison of Methods for the Detection of Bumble Bee Nests," *Journal of Apicultural Research* 51 (2012): 204–11, doi:10.3896/ibra.1.51.2.09.
88. A. Whitelaw et al., "Using Detector Dogs to Locate *Euglandina rosea*" (University of Hawaii at Manoa, Honolulu, 2009); Parker, "Wildlife Detection Dogs."
89. Mussel Dogs, http://musseldogs.info/.
90. C. Warren, *What the Dog Knows: The Science and Wonder of Working Dogs* (Touchstone, 2013); Wildlife Society, http://tws.sclivelearningcenter.com/index.aspx.
91. Conservation Canines, Center for Conservation Biology, University of Washington, http://conservationbiology.net/conservation-canines/.
92. Wagtail, Conservation Dogs, http://www.wagtailuk.com/conservation-dogs/.
93. Browne, "The Use of Dogs to Detect New Zealand Reptile Scents."

94. M. E. Cablk et al., "Olfaction-Based Detection Distance: A Quantitative Analysis of How Far Away Dogs Recognize Tortoise Odor and Follow It to Source," *Sensors* 8 (2008): 2208–22, doi:10.3390/s8042208.
95. F. C. Zwickel, "Use of Dogs in Wildlife Management," in *Wildlife Management Techniques*, ed. R. H. Giles, 3rd ed. (The Wildlife Society, 1969), 319–24.
96. Ibid.; F. C. Zwickel, "Use of Dogs in Wildlife Biology," in *Wildlife Management Techniques*, ed. S. D. Schemnitz (The Wildlife Society, 1980), 531–36.
97. T. Kreeger and J. M. Arnemo, *Handbook of Wildlife Chemical Immobilization*, 3rd ed. (self-published, 2007).
98. P. MacKay et al., "Scat Detection Dogs," in *Noninvasive Survey Methods for Carnivores*, ed. R. A. Long et al. (Island Press, 2008), 183–222.
99. The Wildlife Society, http://tws.sclivelearningcenter.com/index.aspx.
100. R. L. Harrison, "A Comparison of Survey Methods for Detecting Bobcats," *Wildlife Society Bulletin* 34 (2006): 548–52, doi:10.2193/0091-7648(2006)34[548:acosmf]2.0.co;2.
101. Wasser et al., "Scat Detection Dogs in Wildlife Research and Management: Application to Grizzly and Black Bears in the Yellowhead Ecosystem, Alberta, Canada," *Canadian Journal of Zoology/Revue canadienne de zoologie* 82 (2004): 475–92, doi:10.1139/z04-020.
102. Kerley and Salkina, "Using Scent-Matching Dogs to Identify Individual Amur Tigers from Scats."
103. Kauhala and Salonen, "Does a Non-Invasive Method—Latrine Surveys—Reveal Habitat Preferences of Raccoon Dogs and Badgers?"
104. Smith et al., "Relative Abundance of Endangered San Joaquin Kit Foxes (*Vulpes macrotis mutica*) Based on Scat-Detection Dog Surveys."
105. Dahlgren et al., "Use of Dogs in Wildlife Research and Management."
106. Wasser et al., "Scat Detection Dogs in Wildlife Research and Management"; Hurt and Smith, "Conservation Dogs"; C. M. Thompson, J. A. Royle, and J. D. Garner, "A Framework for Inference about Carnivore Density from Unstructured Spatial Sampling of Scat Using Detector Dogs," *Journal of Wildlife Management* 76 (2012): 863–71 doi:10.1002/jwmg.317.
107. Zwickel, "Use of Dogs in Wildlife Management."
108. Ibid.
109. Dahlgren et al., "Use of Dogs in Wildlife Research and Management."
110. S. A. Reindl-Thompson et al., "Efficacy of Scent Dogs in Detecting Black-Footed Ferrets at a Reintroduction Site in South Dakota," *Wildlife Society Bulletin* 34 (2006): 1435–39, doi:10.2193/0091-7648(2006)34[1435:eosdid]2.0.co;2.
111. Grandjean and Haymann, *Hunde-encyklopedi*.
112. C. Lydersen and I. Gjertz, "Studies of the Ringed Seal *Phoca hispida* in Its Breeding Habitat in Kongsfjorden, Svalbard," *Polar Research* 4 (1986): 57–63; C. M. Furgal, S. Innes, and K. M. Kovacs, "Characteristics of Ringed Seal, *Phoca hispida*, Subnivean Structures and Breeding Habitat and Their Effects on Predation," *Cana-*

dian Journal of Zoology/Revue canadienne de zoologie 74 (1996): 858–74, doi:10.1139/z96-100; T. G. Smith, "The Ringed Seal *Phoca hispida* of the Canadian Western Arctic," *Canadian Bulletin of Fisheries and Aquatic Sciences* 216 (1987): 1–81.

CHAPTER TWELVE

1. AP, "Rat-Hunting Dogs Take a Bite out of New York City's Vermin Problem," April 30, 2013, http://www.nydailynews.com/new-york/rat-hunting-dogs-bite-new-york-city-vermin-problem-article-1.1331191.
2. V. Harraca et al., "Smelling Your Way to Food: Can Bed Bugs Use Our Odour?" *Journal of Experimental Biology* 215 (2012): 623–29, doi:10.1242/jeb.065748.
3. M. Goldish, *Pest-Sniffing Dogs* (Bearport Publishing, 2012).
4. Ibid.
5. D. A. Ekeberg, "Midler mot veggdyr sannsynlig dødsårsak på hotel," *Thailands Tidende Nett* 7 (2011).
6. C. Saint Louis, "A Dark and Itchy Night," *New York Times*, December 5, 2012.
7. M. Alomajan, *Dogs in Action: Working Dogs and Their Stories* (Exisle Publishing, 2013).
8. A. Endrestøl, "Veggdyret-tegenes fyrst Dracula," *Insekt-Nytt* 36 (2011): 49–60.
9. Pfiester, Koehler, and Pereira, "Ability of Bed Bug-Detecting Canines to Locate Live Bed Bugs and Viable Bed Bug Eggs."
10. R. Vaidyanathan and M. F. Feldlaufer, "Bed Bug Detection: Current Technologies and Future Directions," *American Journal of Tropical Medicine and Hygiene* 88 (2013): 619–25, doi:10.4269/ajtmh.12-0493.
11. K. F. Raffa, "Terpenes Tell Different Tales at Different Scales: Glimpses into the Chemical Ecology of Conifer—Bark Beetle—Microbial Interactions," *Journal of Chemical Ecology* 40 (2014): 1–20, doi:10.1007/s10886-013-0368-y.
12. F. Schlyter, G. Birgersson, and A. Johansson, "Detection Dogs Recognize Pheromone from Spruce Bark Beetle and Follow It to Source: A New Tool from Chemcial Ecology to Forest Protection" (ESA Sixtieth Annual Meeting, Knoxville, TN, November 12, 2012).
13. W. E. Wallner and T. L. Ellis, "Olfactory Detection of Gypsy Moth Pheromone and Egg Masses by Domestic Canines," *Environmental Entomology* 5 (1976): 183–86.
14. J. B. Welch, "A Detector Dog for Screwworms (Diptera, Calliphoridae)," *Journal of Economic Entomology* 83 (1990): 1932–34.
15. F. Noireau et al., "Can Wild *Triatoma infestans foci* in Bolivia Jeopardize Chagas Disease Control Efforts?" *Trends in Parasitology* 21 (2005): 7–10, doi:10.1016/j.pt.2004.10.007.
16. R. E. Bernstein, "Darwin's Illness: Chagas' Disease Resurgens," *Journal of the Royal Society of Medicine* 77 (1984): 608–9; A. K. Campbell and S. B. Matthews, "Darwin's Illness Revealed," *Postgraduate Medical Journal* 81 (2005): 248–51, doi:10.1136/pgmj.2004.025569; F. Ødegaard, e-mail message to author, October 29, 2012.

17. F. Ødegaard, e-mail message to author, October 29, 2012.
18. M. Rolon et al., "First Report of Colonies of Sylvatic *Triatoma infestans* (Hemiptera: Reduviidae) in the Paraguayan Chaco, Using a Trained Dog," *PLOS Neglected Tropical Diseases* 5 (2011): e1026.
19. K. Warfield, "Man's Best Friend Protects American Agriculture," United States Department of Agriculture, August 12, 2011, https://www.usda.gov/media/blog/2011/08/12/mans-best-friend-protects-american-agriculture.
20. J. Nakash, Y. Osem, and M. Kehat, "A Suggestion to Use Dogs for Detecting Red Palm Weevil (*Rhynchophorus ferrugineus*) Infestation in Date Palms in Israel," *Phytoparasitica* 28 (2000): 153–55, doi:10.1007/bf02981745.
21. A. Russell, "The Nose Knows," *Northbay biz*, February 2007.
22. J. Fjelddalen, "Skjoldlus (Coccinea, Hom.) i Norge," *Insekt-Nytt* 21 (1996): 6–25.
23. Russell, "The Nose Knows"; Fjelddalen, "Skjoldlus (Coccinea, Hom.) i Norge."
24. Russell, "The Nose Knows."
25. V. R. Lewis, C. F. Fouche, and R. L. Lemaster, "Evaluation of Dog-Assisted Searches and Electronic Odor Devices for Detecting the Western Subterranean Termite," *Forest Products Journal* 47 (1997); R. H. Scheffrahn, N. Y. Su, and P. Busey, "Laboratory and Field Evaluations of Selected Chemical Treatments for Control of Drywood Termites (Isoptera: Kalotermitidae)," *Journal of Economic Entomology* 90 (1997): 492–502.
26. Lewis, Fouche, and Lemaster, "Evaluation of Dog-Assisted Searches and Electronic Odor Devices for Detecting the Western Subterranean Termite."
27. S. E. Brooks, F. M. Oi, and P. G. Koehler, "Ability of Canine Termite Detectors to Locate Live Termites and Discriminate Them from Non-Termite Material," *Journal of Economic Entomology* 96 (2003): 1259–66, doi:10.1603/0022-0493-96.4.1259.
28. D. Grandjean and F. Haymann, *Hunde-encyklopedi* (RK Grafisk, 2010).
29. H. M. Lin et al., "Fire Ant-Detecting Canines: A Complementary Method in Detecting Red Imported Fire Ants," *Journal of Economic Entomology* 104 (2011): 225–31, doi:10.1603/ec10298.
30. E. Kauhanen et al., "Validity of Detection of Microbial Growth in Buildings by Trained Dogs," *Environment International* 28 (2002): 153–57, doi:10.1016/s0160-4120(02)00021-1.
31. W. Lorenz and T. Diederich, "How to Find Hidden Microbial Growth with a Mold Dog," *IAQ* (2001): 1–3.
32. A. Ammer and HUSSE (Hundar för Stolprötsdetektering, Säkerhet och Ekonomi), "Dogs Detecting Decay in Telephone Poles, Reliability and Economy," Report (in Swedish), Malmo(1985); L. Davner, "Decay! Dogs Hunting Damaged Wood," *Skogen* 5 (1986): 48; E. Kauhanen et al., "Validity of Detection of Microbial Growth in Buildings by Trained Dogs"; R. T. Griffith et al., "Differentiation of Toxic Molds via Headspace SPME-GC/MS and Canine Detection," *Sensors* 7 (2007): 1496–508, doi:10.3390/s7081496.
33. N. G. Sæther, "Sniffer sopp og sporer mugg," *Hus og Bolig* 5 (2006): 59–63.

34. Davner, "Decay! Dogs Hunting Damaged Wood."
35. S. Järverud, *Din hund söker* (Brukshundservice Sverige AB, 2002).
36. Lorenz and Diederich, "How to Find Hidden Microbial Growth with a Mold Dog."
37. Järverud, *Din hund söker*.
38. Ø. Holen, "Arbeidsulykke blir etterforsket," *TA*, May 15, 2009.
39. T. O. Moen, "En råtten jobb," *Vi Menn*, December 8, 2008.
40. S. Refsnæs et al., "Timing of Wood Pole Replacement Based on Lifetime Estimation" (Ninth International Conference on Probabilistic Methods Applied to Power Systems KTH, Stockholm, Sweden, June 11–15, 2006).
41. Ibid.
42. Moen, "En råtten jobb."
43. P. J. Ann et al., "*Phellinus noxius* Brown Root Rot of Fruit on Ornamental Trees in Taiwan," *American Phytopathological Society* 86 (2002): 820–26.
44. H. Solheim, "Råtesopper- i levende trær," *Skog og landskap* (2010): 1–28.
45. "Sniffer Canines Trained to Find Tree Root Disease," May 25, 2012, http://www.taipeitimes.com/News/taiwan/archives/2012/05/25/2003533696.
46. S. Saha et al., "Survey of Endosymbionts in the *Diaphorina citri* Metagenome and Assembly of a Wolbachia wDi Draft Genome," *PLOS ONE* 7 (2012), doi:10.1371/journal.pone.0050067; T. R. Gottwald, "Current Epidemiological Understanding of Citrus Huanglongbing," *Annual Review of Phytopathology* 48 (2010): 119–39.
47. Center for Invasive Species Research, "Asian Citrus Psyllid," http://cisr.ucr.edu/asian_citrus_psyllid.html.
48. Jon Schärer, "Julestjerne-bakterie påvist I Norge," Forskning.no, December 10, 2010, http://www.forskning.no/artikler/2010/oktober/266754.
49. F. Elfenbein, "Dogs Learning to Sniff Out Citrus Canker," Two Little Cavaliers, December 21, 2011, http://twolittlecavaliers.com/2011/12/dogs-learning-to-sniff-out-citrus-canker.html.
50. C. Martin, "Timber Dogs: A Forest Owner's Best Friend?: *Alabama's Treasured Forest Magazine* (Summer 2011); L. Eckhardt and T. Steury, "Root Diseases and Timber Dogs," *Silviculture* (Fall 2011): 30–31.
51. S. C. de Wet, *Use of a Sniffer Dog in the Detection of American Foulbrood in Beehives*, Australian Government, Rural Industries Research and Development Corporation Publication No. 13/080.
52. K. M. Richards, S. J. Cotton, and R. M. Sandeman, "The Use of Detector Dogs in the Diagnosis of Nematode Infections in Sheep Feces," *Journal of Veterinary Behavior: Clinical Applications and Research* 3 (2008): 25–31, doi:10.1016/j.jveb.2007.10.006.
53. Ibid.
54. S. Alasaad et al., "Sarcoptic-Mange Detector Dogs Used to Identify Infected Animals during Outbreaks in Wildlife," *BMC Veterinary Research* 8 (2012): 110.
55. K. E. Mounsey, J. S. McCarthy, and S. F. Walton, "Scratching the Itch: New Tools to Advance Understanding of Scabies," *Trends in Parasitology* 29 (2013): 35–42, doi:10.1016/j.pt.2012.09.006.

56. R. K. Davidson, S. Bornstein, and K. Handeland, "Long-Term Study of *Sarcoptes scabiei* Infection in Norwegian Red Foxes (*Vulpes vulpes*) Indicating Host/Parasite Adaptation," *Veterinary Parasitology* 156 (2008): 277–83, doi:10.1016/j.vetpar.2008.05.019.
57. Pfiester, Koehler, and Pereira, "Ability of Bed Bug-Detecting Canines to Locate Live Bed Bugs and Viable Bed Bug Eggs."
58. M. M. Bouzga and A. G. Skalleberg, "Hundeprosjektet," *Bevis* 3 (2013): 23–25.
59. L. Gederaas et al., *Fremmede arter i Norge—med norsk svarteliste 2012* (Artsdatabanken, Trondheim, 2012).
60. A. S. Sletmoen and A. Nærheim, "Anmeldt for humleimport," *NRK Hedemark og Oppland*, April 26, 2013.
61. E. Haakaas and W. Fuglehaug, "Maur fra Argentina truer Norge," *Aftenposten*, April 6, 2013; I. L. Stranden, "Ny insektart legger sin elsk på Snillfjord," *NRK Trøndelag*, September 24, 2013.
62. L. Bonesi and S. Palazon, "The American Mink in Europe: Status, Impacts, and Control," *Biological Conservation* 134 (2007): 470–83, doi:10.1016/j.biocon.2006.09.006; K. Bevanger and Ø. Ålbu, *Mink Mustela vison i Norge* (Trondheim: Økoforsk, NAVF, 1986).
63. L. Gederaas et al., *Fremmede arter i Norge—med norsk svarteliste 2012* (Artsdatabanken, Trondheim, 2012); Bonesi and S. Palazon, "The American Mink in Europe."
64. O. I. Edvardsen, e-mail with an unpublished report (2013).
65. B. G. Deshpande, "Earthquakes, Animals and Man, Chapter III: Animal Response to Earthquakes," *Proceedings of the National Academies of Science: India* B52, no. 5 (1986): 585–618.
66. K. I. Aasheim, "Hund som arbeidsredskap i brannvesenet," *Brannmannen* 61 (2006): 42–43.
67. L. R. Quaife, K. J. Moynihan, and D. A. Larson, "A New Pipeline Leak-Location Technique Utilizing a Novel (Patented) Test-Fluid and Trained Domestic Dog," *Proceedings of the Seventy-First GPA Annual Convention* (1991), 154–61.
68. U. Ellervik, *Ond kemi. Berättelser om människor, mord och molekyler* (Fri tanke forlag, 2011).
69. Quaife, Moynihan, and Larson, "A New Pipeline Leak-Location Technique Utilizing a Novel (Patented) Test-Fluid and Trained Domestic Dog."
70. B. Rosén and L. Wetterholm, *Kvalitetssäkringsrutiner för läcksokning av gasledning med konventionella metoder och med sökshundar* (Svensk Gastekniskt Center AB A25, 2000).
71. B. Rosen and L. Wetterholm, *Kvalitetssäkring av läcksøkning med hund-fas II* (Svenskt Gastekninsk Center AB 119, 2001).
72. Järverud, *Din hund söker*.
73. Rosen and Wetterholm, *Kvalitetssäkring av läcksøkning med hund-fas II*.
74. A. Rodrigues, "Hero Dog Alerts Family to Gas Leak," *Toronto Sun*, June 25, 2011, http://www.torontosun.com/2011/06/25/hero-dog-alerts-family-to-gas-leak.

75. A. Schoon et al., "Detecting Corrosion under Insulation (CUI) Using Trained Dogs: A Novel Approach," *Materials Evaluation* 72 (2014): 142–48; E. Tho, "Snuser seg fram til rust," *Haugesund Avis*, March 11, 2013; A. Schoon and R. Fjellanger, *CUI Dog Project Summary Report* (Fjellanger Dog Training Academy, 2013); A. Schoon et al., "Using Dogs to Detect Hidden Corrosion," *Applied Animal Behaviour Science* 153 (2014): 43–52; T. Førde, "Hund får jobb i gassbransjen," *Bergens Tidende*, March 12, 2013.

CHAPTER THIRTEEN

1. J. Murray et al., *Canine Scent and Microbial Source Tracking in Santa Barbara, California* (Water Environment Reseacrh Foundation, IWA Publishing, 2011); S. D. Reynolds, D. P. Christian, and K. G. Reynolds, "Illicit Discharge Is Going to the Dogs," paper presented at the World Environmental and Water Resources Congress 2008, Ahupua'a, Reston, VA.
2. A. Hallgren and M. Hansson, *Kantarellsök med hund* (Jycke-Tryck AB, 2009).
3. W. M. Daniewski et al., "Search for Bioactive Compounds from *Cantharellus cibarius*," *Natural Product Communications* 7 (2012): 917–18.
4. http://www.adressa.no/nyheter/article48831.ece.
5. I. S. Torp, "Trøffelhelter," *Aftenposten. A magasinet*, October 12, 2011.
6. R. Willes, *Lagotto romagnolo* (Mälaröbörsen, 2009).
7. C. Wedén, *Tryffel mat for gudar, gutar & svin* (Infotain & Infobooks, Sweden AB, 2008).
8. D. Grandjean and F. Haymann, *Hunde-encyklopedi* (RK Grafisk, 2010).
9. N. Samils et al., "The Socioeconomic Impact of Truffle Cultivation in Rural Spain," *Economic Botany* 62 (2008): 331–40, doi:10.1007/s12231-008-9030-y.
10. "Putting Tools in Place to Tackle the Threat of Poisoning," in *Life and Human Coexistence with Large Carnivores* (Life Nature, 2013), 63–67.
11. Torp, "Trøffelhelter."
12. http://www.nhm.uio.no/forskning/prosjekter/troffel/metode/.
13. H. L. Odinsen, "Jakten på den sorte diamant," *VG*, December 4, 2011.
14. A. Molia and K. Killingmo, e-mail message to author and conversations during November 2013.
15. T. Talou et al., "Dimethyl Sulphide: The Secret for Black Truffle Hunting by Animals?" *Mycological Research* 94 (1990): 277–78; R. Splivallo et al., "Truffle Volatiles: From Chemical Ecology to Aroma Biosynthesis," *New Phytologist* 189 (2011): 688–99, doi:10.1111/j.1469-8137.2010.03523.x.
16. Wedén, *Tryffel mat for gudar, gutar & svin.*
17. Ibid.; T. Talou et al., "Dimethyl Sulphide: The Secret for Black Truffle Hunting by Animals?"
18. Wedén, *Tryffel mat for gudar, gutar & svin.*
19. Odinsen, "Jakten på den sorte diamant."
20. P. G. Hepper and D. L. Wells, "Perinatal Olfactory Learning in the Domestic Dog," *Chemical Senses* 31 (2006): 207–12, doi:10.1093/chemse/bjj020; D. L. Wells and

P. G. Hepper, "Prenatal Olfactory Learning in the Domestic Dog," *Animal Behaviour* 72 (2006): 681–86, doi:10.1016/j.anbehav.2005.12.008.

21. Wedén, *Tryffel mat for gudar, gutar & svin.*
22. A. Molia and K. Killingmo, e-mail message to author and conversations during November 2013.
23. Grandjean and Haymann, *Hunde-encyklopedi.*
24. M. L. Partyka et al., "Quantifying the Sensitivity of Scent Detection Dogs to Identify Fecal Contamination on Raw Produce," *Journal of Food Protection* 77 (2014): 6–14.
25. R. A. Shelby et al., "Detection of Catfish Off-Flavour Compounds by Trained Dogs," *Aquaculture Research* 35 (2004): 888–92, doi:10.1111/j.1365--2109.2004.01081.x; R. A. Shelby et al., "Detection of Off-Flavour in Channel Catfish (*Ictalurus punctatus* Rafinesque) Fillets by Trained Dogs," *Aquaculture Research* 37 (2006): 299–301, doi:10.1111/j.1365-2109.2005.01416.x.
26. T. R. Hanson, "Economic Impact of Off-Flavor to the U.S. Catfish Industry," in *Off-Flavors in Aquaculture*, ed. A. M. Rimando and K. K. Schrader (American Chemical Society, Washington, DC, 2003), 13–29.
27. R. A. Shelby et al., "Detection of Catfish Off-Flavour Compounds by Trained Dogs."
28. Ibid.; Hanson, "Economic Impact of Off-Flavor to the U.S. Catfish Industry."
29. P. V. Zimba et al., "Confirmation of Catfish, *Ictalurus punctatus* (Rafinesque), Mortality from Microcystis Toxins," *Journal of Fish Diseases* 24 (2001): 41–47, doi:10.1046/j.1365-2761.2001.00273.x.
30. R. A. Shelby et al., "Detection of Catfish Off-Flavour Compounds by Trained Dogs."
31. R. A. Shelby et al., "Detection of Off-Flavour in Channel Catfish (*Ictalurus punctatus* Rafinesque) Fillets by Trained Dogs."
32. P. Chatonnet et al., "Identification and Responsibility of 2,4,6-Tribromoanisole in Musty, Corked Odors in Wine," *Journal of Agricultural and Food Chemistry* 52 (2004): 1255–62, doi:10.1021/jf030632f.
33. A. Stalheim, "Slik avslører du dårlig vin," 2009, http://www.klikk.no/mat/drikke/article397313.ece.
34. S. Gannon, "Winery Dog Sniff Out TCA," *Wines & Vines* (August 2006).
35. Grandjean and Haymann, *Hunde-encyklopedi.*
36. K. VerCauteren et al., "Dogs as Mediators of Conservation Conflicts," in *Free-Ranging Dogs and Wildlife Conservation*, ed. M. E. Gompper (Oxford University Press, 2014), 211–38; J. Beringer et al., "Use of Dogs to Reduce Damage by Deer to a White-Pine Plantation," *Wildlife Society Bulletin* 22 (1994): 627–32; P. Curtis and R. Rieckenberg, "Use of Dogs for Reducing Deer Damage in Apple Orchards," *Proceedings of the Wildlife Damage Management Conference* (2005): 149–58; K. C. VerCauteren et al., "Dogs for Reducing Wildlife Damage to Organic Crops: A Case Study," *Proceedings of the Wildlife Damage Management Conference* (2005): 286–93; R. A. Woodruff and J. S. Green, "Livestock Herding Dogs: A Unique Application for Wildlife Damage Management," *Proceedings of the Great Plains Wildlife*

Damage Control Workshop (1995): 43–45; P. M. Castelli and S. E. Sleggs, "Efficacy of Border Collies to Control Nuisance Canada Geese," *Wildlife Society Bulletin* 28 (2000): 385–92.

37. N. Lescureux and J. D. C. Linnell, "Warring Brothers: The Complex Interactions between Wolves (*Canis lupus*) and Dogs (*Canis familiaris*) in a Conservation Context," *Biological Conservation* 171 (2014): 232–45, doi:http://dx.doi.org/10.1016/j.biocon.2014.01.032.

38. R. Coppinger and L. Coppinger, *Dogs: A Startling New Understanding of Canine Origin, Behavior, and Evolution* (Crosskeys Select Books, 2004); L. L. Marker, A. J. Dickman, and D. W. Macdonald, "Perceived Effectiveness of Livestock-Guarding Dogs Placed on Namibian Farms," *Rangeland Ecology & Management* 58 (2005): 329–36, doi:10.2111/1551-5028(2005)058[0329:peoldp]2.0.co;2; L. L. Marker, A. J. Dickman, and D. W. Macdonald, "Survivorship and Causes of Mortality for Livestock-Guarding Dogs on Namibian Rangeland," *Rangeland Ecology & Management* 58 (2005): 337–43, doi:10.2111/1551-5028(2005)058[0337:sacomf]2.0.co;2; L. Svensson, J. Karlsson, and T. Gustavsson, *Boskapsvaktande hundar i Sverige, viltskadecenters rekommendationer för fortsatt använding* (Rapport från Viltskadecenter Tamdjur/Rovdjur 2011-1, 2011); K. C. VerCauteren et al., "Cow Dogs: Use of Livestock Protection Dogs for Reducing Predation and Transmission of Pathogens from Wildlife to Cattle," *Applied Animal Behaviour Science* 140 (2012): 128–36, doi:10.1016/j.applanim.2012.06.006.

39. M. Weisbord and K. Kachanoff, *Dogs with Jobs* (Pocket Books, 2000).

40. Ibid.

41. C. A. Kiddy et al., "Detection of Estrous-Related Odors in Cows by Trained Dogs," *Biology of Reproduction* 19 (1978): 389–95, doi:10.1095/biolreprod19.2.389; H. W. Hawk, H. H. Conley, and C. A. Kiddy, "Estrous-Related Odors in Milk Detected by Trained Dogs," *Journal of Dairy Science* 67 (1984): 392–97.

42. http://hundcampus.se/; C. Fischer-Tenhagen et al., "Training Dogs on a Scent Platform for Oestrous Detection in Cows," *Applied Animal Behaviour Science* 131 (2011): 63–70, doi:10.1016/j.applanim.2011.01.006.

43. Ibid.

44. C. Fischer-Tenhagen et al., "Training of Dogs for Detection of Oestrous Specific Scent in Saliva of Cows," *Reproduction in Domestic Animals* 46 (2011): 104–5; C. Fischer-Tenhagen, B.-A. Tenhagen, and W. Heuwieser, "Ability of Dogs to Detect Cows in Estrous from Sniffing Saliva Samples," *Journal of Dairy Science* 96 (2013): 1081–84.

45. A. E. Derocher et al., "Rapid Ecosystem Change and Polar Bear Conservation," *Conservation Letters* 6 (2013): 368–75, doi:10.1111/conl.12009.

46. AP, "Meet Elvis the Beagle Who Is Trained to Detect Pregnant Polar Bears," *Daily Mail*, November 4, 2013, http://www.dailymail.co.uk/news/article-2487374/Meet-beagle-trained-detect-pregnant-polar-bears.html.

47. G. Nilsson, "The Use of Dogs in Prospecting for Sulphide Ores," *Geologiska Föreningens i Stockholm Förhandlingar* 93 (1971): 725–28.

48. Grandjean and Haymann, *Hunde-encyklopedi.*
49. R. Austin, *How to Train Your Dog to Find Gold* (Golddogs, 2008).
50. I. Giske, "Jippi—en hund etter gift," *Dagbladet*, January 14, 2007; S. H. Kvalvik, "Hunder som søker etter kreft," *Hundesport* 6–7 (2007): 28–29.
51. H. M. Engdahl et al., "Kartlegging/metodeutvikling av PCB og tungmetaller i fisk og blåskjell fra Idde- og Ringdalsfjorden. 61" (Rapport. BSc thesis. Østfold Ingeniørhøgskole, Sarpsborg, 1991); M. A. Kamrin and R. K. Ringer, "PCB Residues in Mammals: A Review," *Toxicological and Environmental Chemistry* 41 (1994): 63–84, doi:10.1080/02772249409357961.
52. A. Crook, "Use of Odour Detection Dogs in Residue Management Programs," *Asian-Australasian Journal of Animal Sciences* 13 (2000): 219.
53. L. D. Arner, G. R. Johnson, and H. S. Skovronek, "Delineating Toxic Areas by Canine Olfaction," *Journal of Hazardous Materials* 13 (1986): 375–81, doi:10.1016/0304-3894(86)85009-9.
54. K. von Rein and L. D. Hylander, "Experiences from Phasing Out the Use of Mercury in Sweden," *Regional Environmental Change* 1 (2000): 126–34, doi:10.1007/s101130000016.
55. T. W. Clarkson, "Human Toxicology of Mercury," *Journal of Trace Elements in Experimental Medicine* 11 (1998): 303–17.
56. D. W. Boening, "Ecological Effects, Transport, and Fate of Mercury: A General Review," *Chemosphere* 40 (2000): 1335–51, doi:10.1016/s0045-6535(99)00283-0; M. F. Wolfe, S. Schwarzbach, and R. A. Sulaiman, "Effects of Mercury on Wildlife: A Comprehensive Review," *Environmental Toxicology and Chemistry* 17 (1998): 146–60, doi:10.1897/1551-5028(1998)017<0146:eomowa>2.3.co;2.
57. G. M. Lovejoy, "Clancy, the Mercury-Sniffing Dog Stops at St. Jon's School in Wykoff" (2003), http://www.mercuryinschools.uwex.edu/act/clancy.htm.
58. D. Dickins, *Oil in Ice-JIP* (SINTEF Materials and Chemistry, 2010); T. Buvik and P.J. Brandvik, *Using Dogs to Detect Oil Hidden in Beach Sediments: Results from Field Training on Svalbard, September 2008 and on the West Coast of Norway (Fedje/Austreheim), November 2008* (Fedje/Austreheim, June 2009. SINTEF report, Trondheim, Norway, 2009); P. J. Brandvik and T. Buvik, *Using Dogs to Detect Oil Hidden in Snow and Ice: Results from Field Training on Svalbard, April 2008* (SINTEF, 2009); S. Bjøru, "En hund etter olje," *Adressa.no*, March 7, 2009; E. Eriksen and G. Kallestad, "Nese for olje," *Status* 5 (2007): 21.
59. "Putting Tools in Place to Tackle the Threat of Poisoning."
60. I. Fajardo et al., "Use of Specialised Canine Units to Detect Poisoned Baits and Recover Forensic Evidence in Anadalucia (Southern Spain," in *Carbofuran and Wildlife Poisoning: Global Perspectives and Forensic Approaches*, ed. N. Richards and P. Mineau (Wiley, 2011), 147–55.
61. H. Skovronek et al., *Application Opportunities for Canine Olfaction: Equipment Decontamination and Leaking Tanks* (Cooperative Agreement no. CR-812180-01-0.

US Environmental Protection Agency Hazardous Waste Engineering Research Laboratory, Edison, NJ, 1987).

62. "1966: Pickles the Dog and the Football World Cup," *New Light through Old Windows*, July 10, 2010, http://aipetcher.wordpress.com/2010/07/10/1966-pickles-the-dog-and-the-football-world-cup/.

63. E. Thomsen, phone call conversation on September 16, 2013, and e-mail message to author, March 30, 2014.

64. G. Zachrisson, "Hade vi spelet batte på instinkt?" *Golf* 9 (2013): 98.

INDEX

Note: Single names listed below refer to specific dogs. Page numbers followed by *t* indicate a table.